# BIOCHEMICAL STUDIES ON LEUKEMIA

# PROF.DR. SAMI ALMUDHAFFAR

# DR. MOHAMMED SAMI AL-RAICH

# CHAPTER ONE  1

## Introduction

## a- Chromosome abnormalities:

Chromosome abnormalities are very common in Leukemia, and oocur in all major categories of the acute and chronic forms of the disorder. Usually, the abnormality is maintained in a consistent manner in a particular case[6] in keeping with a defect that is replicated with a high degree of fidelity in the clones of cells, arising from the cell in which malignant behaviour first developed[6] [7].

## b- Genetic:

There is a greatly increased incidence of Leukemia in the identical twin of patients with Leukemia. Increased incidence occurs in Down syndrome and other genetic disorders[8].

## c- Immunological:

Immune deficiency states are associated with an increase in hematological malignancy. Occasionally, acute leukemia presents as an a. plastic anemia[3].

## d- Ionizing radiation:

A significant increase in myeloid leukemia followed the atomic bombing of Japanese cities[9]. An increase in leukemia was observed after radiotherapy for ankylosing spondylitis and diagnostic radiographs of the fetus in pregnancy[4].

## e- Cytotoxic drugs:

These particularly alkylating agents, may induce myeloid leukemia, usually after a Latent period of several years[10].

## f- Exposure to benzene in industry:

Benzene and other petroleum derivatives are strongly suspected of being leukaemognic. Benzene has been shown to be a etiologically related to the high incidence of leukemia and aplastic anemia in people occupationally exposed to it[11][12].

## g- Retro viruses:

One rare form of T cell leukemia/ Lymphoma appears to be associated with a retro virus similar to the viruses causing leukemia in cats and cattle[13][14]. DNA sequences of this virus have been detected in the cells of the leukemic T cell clone but not in the normal B cells of affected patients[15].

## 1.3 Patho physiology:

In vitro cultures of leukemia bone marrow's demonstrate abnormal growth patterns[2]. Typical findings are low cloning efficiency, a tendency to abortive colonies, and an increased percentage of light density. The crucial problem is the failure to undergo maturation and produce functional end cells[1].

Leukemic cells (especially in acute cases) may lack enzymes have ✓ altered surface antigens, produce fetal hemoglobin, or perform poorly in functional assay. Since they immature, leukemic cell are likely to stay in the marrow and remain capable of dividing producing, more and more non functional cells that eventually "pack" the marrow[2].

Their presence may inhibit growth cells. As a population, the neoplastic clone has growth advantage over normal cells even though the individual stem cells may actually divide more slowly[3].

Because of ineffective hemotopiesis and resulting high nucleic acid turnover, uric acid production increase. Nephro toxic level may accumulate, because of the tumor cells Lysis caused by chemotherapy and this lysis can also cause hyperkalemia[5]. Marrow replacement by leukemic cells causes anemia thrombocytopenia, and granulocytopenia[3].

Bleeding presents a major threat, but infection is the major killer opportunistic organisms often infect leukemic patients because of diminished immunity[16].

## 1.4 Classification:

On the basis of clinical history and light microscopy, most Leukemia's can easily divided into lymphocytic or myelocytic and acute or chronic and their respective prognoses and therapies established. A group of French , American , and British ( FAB ) hematologists have described a classification system which has achieved a high degree of acceptance (The FAB classification)[17], and is based on blood and marrow morphological features[18], defined by Romovsky and cyto chemical staining[18][19].

Accordingly leukemia is then classified in this system according to the following:

a- Acute

   1- Acute lymphocytic leukemia (ALL).

   2- Acute nonlymphocytic (myelocytic) leukemia ANLL or AML.


b- Chronic

   1- Chronic lymphocytic leukemia (CLL).

   2- Chronic myelocytic leukemia (CML).

## 1.4.1 Methods used in classification of leukemia:

Laboratory techniques may help in distinguishing the major types and sub types of leukemia.

### a- Light microscopy:

Criteria for distinguishing the blast cells of ALL and AML, morphologic subtypes of AML are described in the FAB system. The more mature cells of ALL and CML are usually easily recognized[3] [5].

### b- Electron microscopy:

Electron microscopy occasionally can reveal granules missed by light microscopy. For example peroxidase and acid phosphatase reaction can also be studied at EM level. Peroxidase is positive in the primary granules in blasts of granulocytic series[5].

### c- Histo chemistry:

Histochemical staining brings out pattern useful in differentiating the acute Leukemia's[2]. There are many types of cytochemical tests which are particularly useful for the diagnosis of acute leukemia[1]. Peroxidase or Sudan black, esterase (a non-specific esterase reaction) commonly alpha naphthyl acetate esterase (ANAE)[3]. Lysozyme (nuramidase), periodic acid- schiff (PAS).

### d- Enzymes:

In addition to histo chemical stains, enzyme assays on blood and urine can detect lysozyme (nuramidase) with the same pattern of expression and with levels correlating with the white blood cell count[2].

Terminal deoxynucleotide transferase (T d T), a unique DNA polymerase randomly adds deoxyuncleotide unit to the 3'- OH terminus of DNA with out needing a template[3]. Blasts from 95% of patients with ALL are (TdT) positive, while fewer than 5% of patients with ANLL have blasts positive for (T d T)[3].

### e- Cell surface markers:

Immunologic methods reveal different surface antigens that depend on cell lineage's[5]. Surface immuglobulin (SIg) characterizes all B cells and in a malignant clone, only one light chain type is expressed. The finding of only one light chain types does not prove monoclonality or malignancy, but in a clinical setting it is often interpreted in that way.

### f- Chromosomes:

Presumably the several Leukemia's ultimately derive from the multitude of chromosomal alterations that cause neoplastic transformation[3][5].

### g- Genetic probes:

DNA probes using techniques like the polymerase chain reaction may permit sensitive and specific detection of gene rearrangements[2].

## 1.5 Clinical features:

### a- Sign of marrow depression:

Bruising, bleeding, fever, pallor, and fatigue with shortness of breath related to aspects of marrow can cause bone neutropenia and anemia,

Expanding bone marrow can cause bone pain, also weight loss, night sweat may be present[3].

### b- Sign of organ infiltration:

Splenomegaly and hepatomegaly are commonly found. Lymphadenopathy occurs less often[5].

### c- CNS Involvement:

Central nervous system (CNS) involvement presents in three ways. a-Meningeal leukemia (arachonidal infiltration can produce headache, nausea, vomiting, lethargy, paresis of legs, enuresis, visual disturbances, papilledema, cranial nerve palsies, seizures, and coma).

Leukostatic thrombi (more common in AML) occurs when circulating blasts exceed $100.000/mm^3$ and can lead intracerebral hemorrhage. Subarachnoid hemorrhage due to thrombocytopenia[3][5].

## 1.6 Laboratory features:

### a- Blood:

Anemia present in more than 90% of patients, high white counts may be recongnized[2]. Thrombocytopenia is present, blast may or may not be see in the blood[2].

### b- Bone Marrow:

The bone marrow is largely replaced by Lymphoblasts which account for 50-100% of all marrow cells the above findings are regarded as diagnostic features for case of acute leukemia.

The typical lymphoblast contains around nucleus with diffuse chromatin and one or two nucleoli and scant basophilic cytoplasm with granules[4][5].

In chronic phase the marrow shows marked granulocytic hyperplasia with a predominance of myelocytes, and frequently increased number of megakaryocytes. Bone marrow cells usually resemble those in the blood[2][5][20].

## 1.7 Treatment:

There is no definite treatment for any type of leukemia and the basic concept is how to modify the quality of the patients life which is achieved by a variety of methods, some are designated to treat the complication and improve the symptoms, like treating anemia, infection, other are directed against the leukemic process like chemotherapatics, radio therapy or marrow transplantation[21]-[27].

Recently immunomodulation[5] table (1-1) showsthe mode of action and side effects of drugs and leukemia therapy.

Table (1-1) The mode of action and side effect of drugs and leukemia therapy (nearly all cause, nasea, vomiting, gastrointestinal disturbances and marrow toxicity)[5]

| Drug | Action | Main side effects |
|---|---|---|
| Aklylating agents<br>Cyclophosphamide<br>Chlorambucil<br>Busulphan | Cross- link double-stranded DNA: inhibit RNA formation | Alopecia, haemorrhagic cystitis, cardiac toxicity.<br>Hepatic toxicity, dermatitis.<br>Marrow aplasia, pulmonary wasting syndrome, hyperpigmentation. |
| Corticosteroids | Steroid- receptor complex affects gene expression; reduced prostaglandin and leukotriene formation; direct membrane damage | Peptic ulcer, obesity, hypertension, osteoporosis, diabetes psychosis, hypokalaemia. |
| M-amsacrine | As daunorubicin | Mucositis, hypokalaemia |
| α -Interferon | Induces 2-5 oligo-A-synthetase, activate RNA. breakdown | Flu-like symptoms, fever, thrombocytopenia, leucopenia, cardiac and neurological (at high doses), autoimmune phenomena |
| Antimetabolites<br>Methotrexate | Inhibits dihydrofolate reductase | All antimetabolites: gastrointestinal toxicity, mouth ulcers, myelotoxicity usually with megaloblastic marrow changes, hepatic toxicity (with long-term therapy); renal failure in high doses |
| 6-Mercaptopurine | Inhibits de-novo purine synthesis and purine interconversions | |

| Drug | Action | Main side effects |
|---|---|---|
| 6-Thioguanine | Incorporated into DNA with strand breaks | |
| Cytosine arabinoside Hydroxyurea | Incorporated into DNA with termination of DNA synthesis Inhibits ribonucleotide reductase | Mucositis, cerebellar damage, conjunctivitis with high doses skin atrophy |
| 5-Azacytidine | Inhibits de-novo pyrimidine synthesis; incorporated into DNA and RNA | |
| Vinca alkaloids | Bind tubulin, inhibit microtubule polymerization needed for spindle formation | |
| Vincristine (a) | | Neurotoxicity (peripheral, paralytic ileus, bladder dysfunction); fits (a) > (b) > (c) |
| Vindesine (b) | Mitotic inhibition | Alopecia |
| Vindesine (c) | | Myelotoxicity (c) > (b) > (a) |
| Anthracyclines Daunorubicin Hydroxodaunorubicin (doxorubicin, Adriamycin) Mitoxantrone | Inhibition of topoisomerase 2; free radical generation; DNA intercalation | Mucositis, cardiac damage |
| Semisynthetic anthraquinones Idarubicin Epirubicin | DNA intercalation, stabilizes topoisomerase 2, cross-linked DNA strand breaks | Mucositis, cardiac damage |

## 1.8 Biochemical changes in leukemia:

There are a number of biochemical changes that take place in patients suffering from leukemia, these bio chemical changes are specific in some and non specific in other's. Patients diagnosed as leukemic, characterized by normal serum alkaline phosphatase or slightly raised, lactate dehydrogenase[5]. Serum potassium, may be spuriously raised, due to leakage of intra cellular potassium from platelets or less commonly from lecucocytes after the blood is drawn. In such cases, the potassium level in freshly drawn citrated blood usually normal[2]. The serum vitamin $B_{12}$ and $B_{12}$ Binding capacity are greatly increased due to raised level of transcobalamin 1[5]. The test of liver function are usually moderately abnormal[1]. Hyper calecaemia is present and is then due to bone destruction[3][5].

## 1.9 Tumor Markers:

According to Sell[28] a tumor marker is substance present in cells, tissues, body fluid or produced by a tumor or the tumor's host in response to the tumor's presence, that can be used to differentiate a tumor from normal tissue or to determine the presence of a tumor based on measurement in the blood or secretions[29][30]. It can be measured qualitatively or quantitatively by chemical, immunological, or molecular biological methods[29].

Some tumor markers are specific for single tumor (tumor- specific-markers); most are found with different tumors of the same tissues type (tumor- associated markers). Such as, the prostate specific antigen (PSA) which is produced exclusively by prostate tissue[31].

The tumor markers are present in higher quantities in cancer tissue or blood from cancer patients than in benign tumors or in the blood of normal subjects[29].

Markers produced by cancers include also enzymes, isoenzymes, hormones, oncofetal antigens, carbohydrate epitopes, receptors, oncogen products, Enzymes as tumor markers were suggested Danilw and Sell. M.D[29] were the first group of tumor markers identified such as (LDH), (PSA), (ALP). Their related activities were used to indicate the presence of cancer. Hormones such as tumor markers such as ACTH, HCG, were used for the detection and monitoring of cancer and monitor the treatment of cancer patient[30]. But these compounds (hormones, enzymes.) are still in debate with regard to their role as tumor markers. But the accepted tumor markers are those related to some antigens such as CEA. Many cell surface marker's are glycoproteins or mucins they appear to have better sensitivity and specificity[29].

In general, tumor marker's may be used for diagnosis and prognosis of carcinomas and for monitoring the effect of therapy as well as targets for localization and therapy[28].

## 1.9.1 Clinical application of tumor markers:

The potential uses of tumor markers are summarized in table (1-2). In general tumor markers may be used for diagnosis and prognosis of carcinomas and for monitoring effects of therapy as well as targets for localization and therapy[32]. Ideally, a tumor marker could be produced by tumor cells be debatable in body fluids.

It could not be present in healthy people or in benign conditions. Most tumor markers present in normal, benign, and cancer tissue, and are not specific enough to be used for screening cancer[32].

## Table (1-2)

### Potential uses of tumor markers

| |
|---|
| * Screening in general population |
| * Differential diagnosis in symptomatic patients |
| * Clinical staging of cancer |
| * Estimating tumor volume |
| * Prognostic indicator for disease progression |
| * Evacuating the success of treatment |
| * Detecting the recurrence of cancer |
| * Monitoring responses to therapy |
| * Radio immuno locallization of tumor masses |
| * Determining direction for immunotherapy |

The marker value at the time of diagnosis may be used as a prognostic indicator for disease progression and patient survival. This is possible for an individual patient, but different levels of markers produced by different tumors do not usually allow one to determine the prognosis of a tumor from the initial level. However, after successful initial treatment, such as surgery, the marker value should be decreased. The magnitude of marker reduction may reflect the degree of success of the treatment or the extent of disease involvement[29].

Most tumor marker values correlate with the effectiveness of treatment and responses to therapy, for example, in breast cancer, markers such as CA-15-3 and CA-549 change with the treatment and the clinical out come of patient. Marker values usually increase with progressive disease decrease with remission, and do not change significantly with stable disease[33].

The marker values in response to treatment may show an initial delay before demonstrating the expected pattern of change[33]. In addition, antibodies to tumor markers labeled with radioactive tag are used to localize the tumor masses, or to provide direction for labeled antibodies to attack the tumor site[34], such as the use of radiolabeled antibodies to CEA to localize colonic tumors and the application of labeled antibodies against ferritin to target hepato cellular carcinoma[34].

### 1.9.2 Methods for evaluating tumor markers:

To evaluate the usefulness of a tumor marker, it is necessary to establish reference values, calculate predictive values, evaluate the distribution of marker value and determine these values role in disease management.

#### a- Reference values:

Reference values of tumor marker are usually established in a healthy population, preferably using age- and sex- matched individuals[35]. The reference values determined using healthy subjects in this fashion is applicable to analytes with physiologically well-defined concentrations, using benign patients as the non disease group is more appropriate than using a healthy population.

#### b- Predictive value model:

The predictive value model includes sensitivity, specificity, and the predictive value of a test. By varying the decision level, sensitivity and specificity will change in opposite directions[35].

## c- Distribution of marker values:

Most tumor markers are elevated in more than one disease condition. The distribution of tumor marker values is usually expressed as the percentage of patients with elevated values as determined using various "cut-off" values in the healthy, benign, and cancerous groups[29].

## 1.9.3 Tumor marker in leukemia:

A large number of suggested biochemical and immunelogical tumor markers has been described, but few of them used in diagnosis, monitoring leukemia[36]-[38], such as LDH, Regan isoenzyme, SOD, TSA, LASA seromucoid proteins, hexoses...etc.[39][40].

A significance increase in serum LDH levels of hematological malignancies has been documented[41][42]. Even though regan isoenzyme is a less specific onco development marker, its value as potential tumor marker has been advocated[43][44].

On the other hand, a decreased of activity of the super oxide dismutase (SOD) has been found in leukemia and all malignant tumor investigated[45]-[48]. Elevated levels of several glycoconjugates including different forms of sialic acid (SA) have been reported in the plasma/ serum of cancer patients[49] colon[5], lung[51], bladder[52], ovarian[53], cervical tumor of CNS[55], prostate[56], myloma[57], skin, kidney[58], Breast[59], leukemia[39][60].

An elevation in the serum concentration of total sialic acid (TSA) has been reported in the majority of children with leukemia[61]. Elevation in serum (TSA) has also been reported in adults with acute myeloid leukemia (AML) and chronic myeloid leukemia[62], acute lympho blastic leukemia(ALL)[39], chronic lymphocytic leukemia[63], and lymphoma[64][65], elevated TSA/ TP values have also been reported in the AML/ CML and ALL patients[66]. Also

(LASA) lipid associated sialic acid elevated in various malignancies including Leukemia's and lymphoma[60][62][67]-[75].

Determination of (LASA) in the extract serum has been reported as a more specific tumor marker than (TSA) and may be more useful than the assay of total sialic acid (TSA)[60][76][77].

Recently (TSA) and (LASA) used in staging and monitoring treatment of leukemia and lymphoma[78][79][80]. Although (SA) binding to glycoprotein is used for monitoring therapy in leukemia and lymphoma patients[81][82].

Sinha. et al.[83] identificated sialic acid as a biomarker in (ALL) using a lectin as a probe. Sinha. et al(84) developed a simple blood assay using (SA) for monitoring chemotherapy for (ALL) patients.

## 1.10 Sialic acid:

Sialic acid, a family of acetylated derivatives of neuraminic acid, is widely distributed in mammals. It is usually occurs as a terminal component at non-reducing end of carbohydrate chains of glycoproteins and glycolipids. Sialic acid is the common name for acylated derivatives of neuraminic acid, a nine- carbon poly hydroxy amino- keto acid sugar (5- amino – 3,5- dideoxy-D- glycero- D- galactononulosonic acid) are widely distributed nature[86]. Figure (1-1).

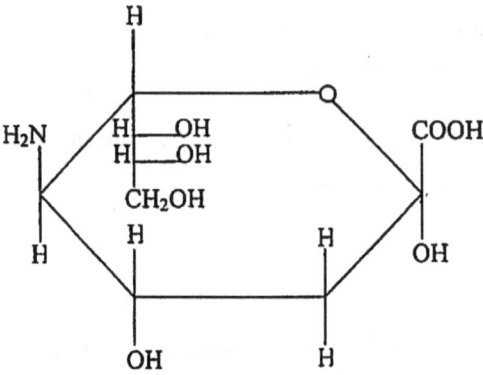

D-Neuraminic acid

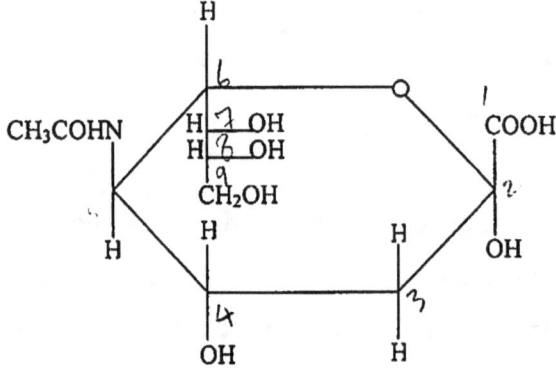

N-Acetyl Neuraminic acid

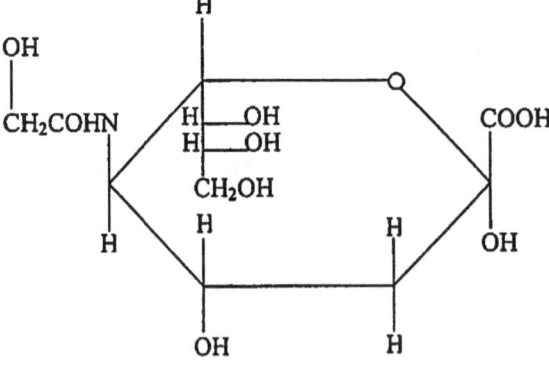

N-glycolyl Neuraminic acid

**Figure. (1.1): Structure of the Sialic acids**

The most commonly occurring of these is N- acetyl neuraminic acid (Neu SAC), which is an important constituent of the oligosaccharide shains found in the glycoproteins and glycolipids of the cell coats and membranes of animal tissue[87].

Some of these sialic acids are known to carry 0-acetyl substiuents located to carbon No.4 and / or at the various position of the polyhydroxy side chain which are carbon No.7, 8 and 9[88][89], Figure (1-2).

Sialic acids (SA), carboxylated mono saccharides often found as terminal residues of vertebrate oligo saccharid[90]. They are modified in various ways giving rise to a family of more than 30 different structures which in turn can be presented in a variety of linkages to the underlying sugar chains[91][92]. In animals, sialic acid are found in α-glycoside linkage as the polysialolyl linkage, and certain other rare exceptions[93]. Sialic acid also have been found in prokaryotic cells as a constituent capsular polysaccharide of a few genera of pathogenic bacteria[94].

C₆ = unsubstituted sialic acids

**Figure. (1.2): Structures of 0- acetyl substitute sialic acids**

## 1.10.1 Sialic acid in tissues and body fluids:

Sialic acids are present in higher animals and certain bacteria but not plants or lower invertebrates[95]. Within mamalian species the occurrence of sialic acids varies considerably [8][96].

In human plasma a large quantity of sialic acid is found in orsomucoid, alpha-1-antitrypsin, haptoglobin, ceruloplasmin, fibrinogen, complement proteins, and transferring[97]. Also, it was purified from cervical mucus, ovarian cyst, lipid- free fraction of brain, and liver[98].

Recently D., Sinha et al., identified two (9-0 acetyl) sialoglyconjugates on lymphoblastic of children suffering from (ALL), as a biomarker using lectin as a probe[83]. Sialic acids (NeuAc), a family of 9- carbon carboxylated monosaccharides[90], are important constituents of lymphocyte cell membrane[99]. Amongst the 0- acetylated derivatives, although 9-0 acetylatecd sialic (9-OACSA) are detected at low level on normal human B lymphocytes[100].

## 1.10.2 Physiological Role of silaic acid:

The human red blood cell is studied with nearly 20 million molecules of sialic acid on the outer cell membrance which contributes to its electronegative change (Zeta potential), and by cell to cell repulsion, prevents red blood cells from aggregating. Owing to it negative charge, sialic acid can bind positively charged molecules (carboxyl group of sialic acid is always free and because its pka of 2.6 is negatively charged at physiological pH) and thus play a role in the transport of such molecules[101].

Since they are essential component of cell surface receptors for a wide range of endogenous (such as peptide hormones) and exogenous (bacteria and viruses) substances, their presence is both a blessing as well as a scourge[102].

Thus, they have a role in the cellular actions of hormones, such as insulin, and also can modulate amino acid transport in some cells. On the other hand, interfection by bacteria or virus is solely dependent on the presence of sialic as a component of specific receptors for the micro organisms on the cell membrane. As antigenic determinates of glycoproteins or glycolipids, sialic acid molecules contribute to the specific of blood group substance[104]-[106].

The negative charges on sialic acid have an influence on the conformation of glycoproteins in terms of their proper alignment in cell membranes, the expression of enzymatic activity of glycoprotein enzymes, and even resistance to proteolytic enzyme degradation[106].

Indeed the clustering of cell membrane glycoproteins owing partly to the repulsion of their oligosaccharide sialic acid residues, is important for cell rigidity, since the loss of sialic acid molecules can increase the deformability of cells[107]. An intriguing role for sialic acid is in its ability to serve as biological masks by preventing ligands from recognizing receptors. Thus a glycoprotein layer rich in sialic acid act as an immune barrier between the mother and the fetus.

Indeed, this masking effect is lost by the removal of terminal sialic acid residues from oligosaccharide chains since it leads to the exposure of a penultimate galactose residue which is then recognized and bound by naturally occurring antibodies, thus facilitating the removal of glycoprotein or the cell by the reticulo endothelial system[105].

They are mainly terminal in oligosaccharide chains and participate in molecular interactions[106]. Sialic acid are modified in various ways and presented in different linkages to the underlying sugar chains which significantly effect their diverse biological function[107]. Modifications of cell surface sialyl residues are associated with lymphoblastogenesis, and lymphoproliferation and their 0- acetyation is reported to alter the immunopotency of sialoglyco conjugates[108].

The modifications of sialic acids significantly effect the physiochemical properties of parent molecule (Neu 5 Ac/ SA) and can there by modify and/ or create new biological function[107]. In mammals, one of the most common modification, of sialic acids in the addition of 0- acetyl esters to hydroxyl groups at $C_4$, $C_7$, $C_8$ and $C_9$ positions. The 0-acetylation of sialic acids influences enzymatic reactions in the catabolism of glycoconjugates[108], effects recognition of sialic acids by viral hemagglutininus and bacterial sialidases[91][92], may effect tissue morphogensis during development[109] and can modulate the alternative path way of complement activation[110].

## 1.10.3 Relevance of sialic acid to tumor:

The relevance of sialic acid to the tumor cell is apparent from the increased sialylation and sialytransferase activity observed in many cancer cells[98]. The aberrant glycosiaylation found in cancer cell membranes is presumably due to the activation of new glycosyl transferases that are characteristic of tumor cells and are absent or present only in small quantities in normal cells[108].

Thus for instance, relatively specific sialyl transferees is found to be present by as much as 2.5 to 11 times in greater amount in transformed cells when compared to control cells[112].

Although the major sialic acid in human is Neu 5Ac, 9-0 acetylated sialic acids (9-0 Ac SA) have been detected at a low level in human saliva, colonic mucins, serum and B lymphocytes[100], and are markedly elevated in melanomas[113], and ALL[114][115]. However, the detailed biochemical characterization and biological significance of g- OAc SA on lymphoblasts of ALL patients remain obscure.

## 1.11 Methods for determination of sialic acid:

A variety of procedures have been used for the measurement of total sialic acid. These can be broadly classified as colorimetric, fluorometric enzymatic, then highly sensitive high performance liquid chromatographic (HPLC) procedure.

### 1.11.1 Colorimetric procedures:

Two classical procedures have stood the test of time one uses resorcinol and the other uses periodic and thiobarbituric acids[116].

#### a- Resorcinol method:

The resorcinol based assay uses heat and strong acid to hydrolyze glycosidic bonds. The released free sialic acids are reacted with resorcinol and copper ions to give colored compound which is extracted and measured at 580nm.. To overcome interference from sugars forming furfural and furfural analogues, such as pentoses, and other interference, such as glucuronic acid and 2- deoxyglucose which incidentally have an absorption maxima at 450nm., and second maxima at 580nm., measurement are also taken at 450nm. Sialic acid concentration is then calculated using simultaneous equation[116].

#### b- The thiobarbituric acid assay:

This procedure described by Warren, which measure only free sialic acid that is released after an initial hydrolysis step. In this procedure, formyl pyruvic acid formed as a result of periodic acid oxidation of free sialic acid is reacted with thiobarbituric acid to yield a red color which is measured at 549nm.

Interference owing to malonaldehyde, the oxidation product of 2-deoxyglucose, is corrected by taking measurements also at 532nm.[117].

### 1.11.2 Fluorometric procedures:

In a typical and more specific assay formaldehyde that is formed upon oxidation of free sialic acid by periodic acid is reacted with acetyl acetone. The yellow product is excited at 410 nm, and the resulting fluorescence is measured at 510nm.[188].

### 1.11.3 Enzymatic procedures:

Enzymatic assay is based on conversion of free sialic acids released by the enzyme neuraminidase to pyruvate and N- acetyl mannosamine with the aid of the enzyme acetyl neuraminic acid pyruvate lyase or NANA aldolase[119].

### 1.11.4 High performance liquid chromatographic procedures:

The HPLC procedures provide the ultimate sensitivity. In one such procedure, sialic acid released from the sample by acid hydrolysis is converted to highly fluorescent derivatives by reacting with fluorogenic agent for alphaketo acids such as 1,2-diamino 4,5-methylene dioxy benzene in dilute sulfuric acid. The fluorescent derivatives are separated on an octadecyl (C18).bonded sialic column using a reverse phase solvent system. The chromatographic step taken only 12 minutes allowing detection of levels as low as 7.7 Pico grams (pg) of N-acetyl neuraminic acid and 7.5 pg of N-glycolyl neuraminic acid, in an injection volume as small as 10 micro liters. The procedure is capable of analyzing precisely sialic acid in a 5 µl serum sample[120].

## 1.12 Sialic acid concentration in different stages of malignant diseases:

Malignant diseases are process encompassing the whole body: they disturb the structure and function of many organs and system. One of these is the change in the composition of serum proteins, including a decrease in the content of albumin with a concomitant increase of glycoproteins that include, among other N-acetyloneuraminic acid or silaic acid. The sialic acid in blood serum is derived mainly from liver cells where it is synthesized from glycolipids and glycoproteins of the membranes of lysed cells[121]. Investigation done by Bosman and Hall show that the enzymes degarding cell membranes are active in cancer cell[122].

The activity of sialylotransferase in the homogenate of cancer cells was 2-7 times higher than in normal tissue. As the determination of sialic acid is simple than the determination of sialylotransferase activity, it has become the subject of interest as a malignancy marker[67][123].

At the same time it was noted that cancer cells contain almost twice time the amount of sialic acid in their cell membranes as normal cells. Many studies have attest to the raised concentration of sialic acid in serum from patients suffering from various neoplastic diseases. It has been considered a useful neoplastic marker, sensitive but non- specific[124].

Comparing the concentration of sialic in different staging of the diseases, they conclude that the concentration of sialic acid increasesin serum blood of patients with neoplastic diseases, and its determination may be useful in monitoring the neoplasmic process in patients with hematological malignancies[64][79][82].

Although decreasing during the treatment and although seem that determination of sialic acid concentration may be useful in monitoring treatment[61][63][125].

## 1.13 Lipid- associated sialic acid (LASA) levels in cancers:

Dnistrian and schwartz[82] were the first, in 1983, to begin determination of plasma lipid- associated sialic acid (LASA) as a sensitive marker in leukemia and malignant lymphoma they have reported raised level of LASA in cancer patient[82].

### 1.13.1 Measurement of lipid- associated sialic acid (LASA):

The basis for the measurement of LASA in serum is the increased level of gangliosides seen in patient in various type of cancers. Apparently, these gangliosides are shed from the tumor cell surface; since they have a relatively long half- life, compared to lipid lacking sialic acid, their accumulation in serum lends it self to its measurement[126].

However, there is a difference in levels of LASA depending on the procedure used for measurement. This is because some procedures measure predominantly glycoprotein- bound sialic, thus grossly overestimating the LBSA. Essentially, these procedures involve extraction of glyco lipid- bound sialic acid with solvents such as chloroform- methanol and the gangliosides are separated from other lipids either by phase partition or precipitated with a one percent phosphotungstic acid solution. The isolated LASA fraction is measured by either the classical resorcinol or periodic- thiobarbituric acid method[129].

## 1.14 Usefulness of TSA & LASA for monitoring:

In typical study, four groups of cancer patients were evaluated, these groups included 69 patients with bladder cancer, 58 patients with lung cancer, 31 patient with cancer of the uterus, and 29 patients with breast cancer. In addition TSA and LASA levels in sera, the carcinoembryonic antigen (CEA) levels were also measured. The sensitivity of the three aasay prior to initiation of radiotherapy, in relation to established cut- off values given parenthesis were: TSA 89.3% (80 mg/dL), LASA 88.8% (20 mg/dL), and CEA 26.8% (5mg/ml). After completion of radiotherapy, the overall response to treatment as judged by percent of patients with a final serum level below the zero time value for each of these marker were: LASA 85.6%, TSA 81.3% and CEA 65.8%. Thus, in this study, the diagnostic sensitivity of LASA and TSA terms of their ability to detect true positives was more than three times greater than CEA. For monitoring response to treatment, TSA and LASA were able to follow correctly 15% more patients with post treatment values below the zero time values when compared to CEA[128][129].

## 1.15 Glycoproteins:

Glycoproteins are an conjugated proteins containing as prosthetic group (one or more) heterosaccharide (S), which include hexoses (galactose and mannose), sialic acid, methylpentose (fucose) and amino sugars (N-acetylglucosamine[130], and N- acetylalactosamine) and (S) is usually branched, lacking repeating units, and bound covalently to the peptide chain[131].

There are great variation in chemical and physical properties of glycoporteins according to their location and function[129] Glycoproteins are isolated from most organisms including: plants, bacteria, fungi, viruses and animals, and their chemical structure differ in the different organs, furthermore the chemical structure of glycoproteins may differ even in different parts of the same organ in the same individual[133].

## 1.15.1 Functions of glycoproteins:

Glycoproteins may serve in many functions, table (1-2), glycoproteins may serve as a structural molecules in the cell, the major protein of the glycoprotein in the animal cell is associated with cell surface, and approximately 70% of the sialic acid- containing glycoproteins are found in the surface membrane[134].

Glycoproteins participate in a large number of normal and disease-related function of clinical relevance. For example many of the proteins in the outer cellular membranes are glycoproteins. Such as proteins within the cell surface are antigens, which determine the blood antigen system, (A,B,O), and the histo- compatibility and transplantation determinates of an individual. Immunoglobulin antigentic sites and viral and hormone receptor sites in cellular membranes are of the glycoproteins[133].

In addition, the carbohydrate portions of glycoproteins in the membranes provide a surface code for cellular identification by other cells and for contact inhibition in the regulation of cell growth. As such, changes in the membrance glycoproteins can be correlated with tumorigenesis and malignant transformation in cancer. Most of the important plasma proteins, except albumin, are glycoproteins[133].

These plasma proteins include the blood- clotting proteins, the immuno- globulins, and many of the complement proteins[133]. Some of the protein hormnes, such as follicle- stimulating hormone (FSH) and thyroid-stimulating hormone TSH, are glycoproteins[133].

The important structural protein collagen contains carbohydrate. The proteins found in mucus secretions are carbohydrate- containing proteins, where they protection of epithelial tissue.

The antiviral protein interferon is a glycoprotein[135]. Glycopteins have other function such as lubricant and protective function, and it serve as a transport molecules for vitamin, lipid, minerals, trace elements and hormones[136].

Lectin are also glycoproteins, the ability of lectins to interact with soluble glycoproteins has been use to isolate and fractionate glycoproteins. This ability is due to binding of lectin to the oligosaccharides moieties of glycoprotein[136].

## 1.16 Lectin:

Lectin is a general term to haemo- agglutinating substances present in extracts of seeds of certain plants, which specifically agglutinate RBC of certain blood group.

Lectins are multivalent specific carbohydrate- binding proteins or glycoproteins of non- immune origin which are widely found in nature including plants and vertebrate tissue[137]. Binding of lectins to the carbohydrate moieties of glycoproteins on the surface of cells has been shown to be involved in variety of biological recognition process[138].

As well as the ability to agglutinate red blood cells, which makes easy their detection, lectin exhibitsa host of other interesting and unusual biological and chemical properties[139], table (1-3). The interactions of lectins with cells can, in many instances,be inhibited specifically by simple sugars.

This finding has led to conclusion that lectins bind specifically to saccharides on the surface of cells, and has provided a new tool for the investigation of the architecture of cell surface. Lectins also bind mono- and oligosaccharides and specifically precipitate polysaccharides and glyoproteins the precipitation inhibited by sugars[140].

Lectins are classified into small number according to their specificity groups (mannose, galactose, N-acetyl glucosamine, N- acetyl galactosamine, L- fucose, and N- acetyl neuraminic acid) and to the mono saccharide that is the most effective inhibitor of the agglutination of erythrocytes or precipitation of carbohydrate- containing polymers by the lectin[140][141].

The specificities of variety of lectins towards N-linked oligosaccharides have proposed as a good tool in the fractionation and structural assessment of oligosaccharides and glycoproteins[142][143].

## 1.16.1 Sialic acid –binding lectins:

Many lectins which have been purified and characterized but few bind sialic acid[136][144]. Wheat germ agglutinin (WGA) is the only plant lectin that binds to sialic acid[145].

Two lectins specific for sialic acids have been purified from animal sources[146], and recently another two lectins, specific- sialic acid have been purified[147][148]. Lectins are found in all classes and sub classes of invertebrates[149] almost in the hemolymph and sexual organs[150].

Lectins also present on the membranes of hemocytes cells[151], that function as primitive and rather unspecific immunological receptors. Few of these lectins have been purified and characterized, most of them are specific for sialic acid[151].

## 1.16.2 Biological functions of lectins:

It is reasonable to expect that the known biochemical properties of lectins dictate their endogenous biological function[152]. One major property of lectins is their specific saccharide- binding sites. In the lectins that are multimers of an identical subunit these binding sites are the same. In contrast, some lectins are composed of subunits with different binding sites. These include the lectin from the red kidney bean, phaseolus vulgaris. It is composed of two different sub units combined into five different forms of noncovalenty bound tetramers[153]. Since the subunit have markedly different specificties for cell surface receptors, each combination could be envisioned to have a different functions. For example, the homo tetramer that contains only these sub units per molecule might inhibit such agglutination[154].

The common finding of more than one lectin in seeds, slime molds and even vertebrate tissues[155], raises the possibility of concerted specific reactions due to concurrent display of these proteins on a structure like cell surface[156].

The marked abundance of lectin, suggests that they play a structural role rather than an enzymatic or catalytic function. It, of course, remains possible that some functions of lectins may be mediated by properties that have not been discovered[157].

Neoplastic transformation causes changes in cell surface architecture, most notably, aberrant sialyation. Exploiting the restricted specificity of a 9-0 acetyl sialic acid (9-0 Ac SA) binding lectin, $Achatinin_H$ ($ATN_H$), have identifies two 9-0 acetylsialoglyconjugates (9-0 $AcSG_S$) on lymphoplasts of children suffering from acute lymphoblaststic leukemia (ALL)[83].

Lectin or lectin- like proteins endowed with ability to bind sialoglycoconjugates have been used as recognition molecules to predict changes in the pattern of sialylation and the degree of 0- acetylation during malignant transforming[106][158][159][160]. Binding studies with sambuccus nigra agglutinin (SNA) state de novo expression of specific sialic acids on selected glycoproteins in colon carcinoma[78]. Employing Ricinus cummunis agglutinin, the granulocytes of patients with chronic myelogenous leukemia

CML are reported to contain more sialylated glycopeptides than normal granulocytes[161][162]. Lymphoblasts of children with ALL are also reported to be electrophoretically distinct from normal lymphocytes due to their increased concentration of sialic acid which causes charge differences[163]. A lectin cancer antennarius, which recognizes sialic acids 0-acetylated both at $C_4$ and $C_9$ positions, has been used to identify an 0-acetyl ganglio side, $CD_3$, as a biomarker in human melanoma cells[113].

### 1.16.3 Lectins applications:

Lectins play an important role in research into variety of cellular properties and processes[164]. Their major biological effects, such as cell agglutination and mitogenic stimulation, appear to be mediated initially through interactions at the level of cell surface and mimic various physiologically important processes. Since lectins can be obtained in a purified form and show well-defined interaction specificities, they have frequently been used in model systems. Boldt coworkers, used lentil lectin, to study the mitogenic responses of various populations of human lymphocytes[165].

Lectins also provide convenient "marker" in cell surface studies. Lentil lectin was used by Scott Rosenthal[166], to characterize plasma membrane vesicles shed guinea pig macrophages on exposure to sulphydryl- blocking reagents.

Table (1-3)

## Some common properties of lectins that suggest biological functions

| Property | Function suggested |
|---|---|
| 1. Specific binding sites:<br>a. All of one kind<br>b. Of different kinds | Recognition of complementary oligosaccharide receptors (range of specificities) |
| 2. More than one carbohydrate binding site | a. Cross-linking glycoproteins or glycolipids in membranes and/ or solution<br>b. High affinity (multisite) binding to molecules or a cell surface with multiple receptors |
| 3. Agglutinin | Binding cells together:<br>a. Like cells (promoting adhesion, fusion, etc.)<br>b. Unlike cells (promoting symbiosis, infection, phagocytosis, etc.) |
| 4. Abundant | Structural rather than catalytic function |
| 5. Generally not integrated in membranes | Relative freedom of movement in or between cellular compartments |

# CHAPTER TWO

# 2

## Total and lipid - associated sialic acids levels in sera of Leukemic and Hodgkin's patients

A  sensitive and specific blood test for early detection and subsequently for the management of cancer patients would be of great clinical value, however, the quest for such a test is still ongoing[66]. Increased levels of enzymes, glycoproteins, hormones, oncofoetal proteins and peptides have been considered as potential tumor markers  for helping in screening, diagnosis, staging, prognosis and monitoring of cancer treatment[66][71][167].

Alterations  in the cell surface during transformation of normal cell to a malignant are well known[168]. Aletrations in serum plasma  levels of total sialic acid (TSA) and lipid associated sialic acid (LASA) levels have been reported to be useful in early diagnosis of cancer as well as in the management of patients with various malignancies[113][169][170].

Studies of cancer cells have revealed alteration in cell  surface and change in plasma  membranes  in form of sialic acid content of glycoproteins and glycolipids as possible markers for these  cancer cells producing them[171][174]. Other studies  of serum or plasma total or lipid associated sialic acid content has been  studied in patients with malignant melanomas[175][176], carcinoma of the ovary, colorectal cancer[177], breast cancer and other cancers, and leukemia[178][179]. Several glycosyltransferase levels have been reported to be abnormal in human cancer[180][183].

Serum sialic acid and sialyltransferase have used  to monitor cancer burden in cancer melanoma patients[184] and  it appears to be useful in this case[79][168].

Some authors have observed an increased level of sialic acid containing glycoconjugates  in sera of malignancies patients. These investigators reported significantly  elevated  serum concentrations  of total sialic acid (TSA), TSA/

total protein (TSA/ TP) and lipid Associated sialic acid LASA at diagnosis of of children with various malignancies compared to healthy children[61][115]. In neuroblastoma and yolk sac tumors, a descending trend in TSA/TP was noted after successful treatment of malignancy patients with simultaneous malignancy[60][80].

Some authors assayed LASA in normal volunteers, patients with non malignant disease and variety of cancer patients (breast[158] Lung, colon, ovarian, prostate[186][187], thyroid, pancreas, adrenal cancer[187] and leukemia)[39][40]. Elevated levels of several forms of sialic acid (SA) have been reported in the plasma/ serum of leukemic patients[39][40], a significantly elevated in (SA) levels of leukemia patients compared with normal and anemia patients[39]. Patel, et al., study serum levels of TSA/TP, LASA, TSA. They found that TSA/TP and LASA to be the most useful, among the markers tests for detecting leukemia[66][40]. Tomaszewska, et al., determined the sialic acid (SA) concentration in the blood serum of children suffering from leukemia and malignant lymphomas. A significant light concentration of (SA) was found at the onset of the disease, compared to controls. In the children, suffering from leukemia (ALL) a dependence was noted between the stage of the disease and the concentration of sialic acid in the serum. The high concentration of sialic acid after onset of the disease, although decreasing during treatment. It seems that determination of sialic acid (SA) concentration my be useful in monitoring treatment[79].

Other authors observed a significantly elevated in serum (TSA) total sialic acid LASA and TSA/TP. TSA and TSA/TP were superior to LASA in differentiating solid tumors and Leukemia's from normal and controls. Also a descending trend in TSA, TSA/TP was noted after a successful treatment of the malignancy, on the other hand, children with (ALL) and infection had a significantly higher serum TSA and TSA/TP levels compared with children with infections disease only[80]. Recently Sinha, et al., used a 9-0 acety sialic acid binding lectin to determine treatment and predict relapse in children with ALL[84].

# Chemicals, Instruments, and Patients samples

## 2.1 chemicals:

All common laboratory chemicals or reagents were of analar grade or the equivalent, and were used with out further purification. They were obtained from the following companies:

a- Na,K – tartarate, Folin- ciocalteau, Chloroform, Butyl acetate, HCl, Perhloric acid, $H_2SO_4$ Benzoic acid, Mannose, Fructose, NBT, Riboflavin, EDTA NaCl, NaOH, $CaCl_2$, KCN, Neuraminidase, $ZnSO_4$, $MgCl_2$, NaF, KCl, $MnCl_2$, NaI, KI. $ZnCl_2$.Sodium pyruvate from BDH company.

b- Ethanol, Methanol , Galactose, Oricinol, Tris- (hydroxy methly amino methan), Bovine serum albumin (BSA) and NADH from Fluka company .

c- Resrcinol, $Na_2HPO_4$ and Urea from May and Baker company.

d- Phosphotungstic acid, $CaCO_3$, $CuSO_4$, $5H_2O$, $KH_2PO_4$ and Xylose from Riedel- De Haene company.

e- Poly ethylene glycol from Merck company.

f- Sialic acid and D- Glucuronic acid from Sigma company.

## 2.2 Instruments:

a- LKB spectrophotometer ultra spec type 4050.

b- Pye- unicam pH meter.

c- MSE centerifuge .

d- Memmert water bath.

e- SM – shaker.

## 2.3 Patients samples:-

Thirty-seven samples of blood taken from non-malignant patients, these samples were used as a pathological control. The age of patients (control) between (20-60) years, and consist of 15 females and 22 males, who gave no history of previous diseases. 12 normal healthy donors.

One hundred ninety samples of blood were taken from patients with blood diseases. Depending upon hematological and histopathological finding, a final diagnosis . consisted of the presence of malignant plasma cells in bone marrow, clinical features and blood film. They were classified into four groups depending on the morphology of the blast cells in bone marrow (lymphocytic or myleocytic), group 1, 152 patients with known leukemia (ALL, CLL, AML, CML), 26 with Hodgkin's disease, all patients were admitted for diagnosis and treatment to Saddam Medical Hospital under the supervision of Dr. Hassan Ibraheem. Then the patients under went therapy. The diagnosis of acute leukemia is based on the presence of 50% blast cells in the bone marrow. In patients with high levels of circulating white blood cells (WBC) e.g $(50 \times 10^9/1)$, the diagnosis is self- evident from the peripheral blood. The host information of patients and control person are summarized in table (2-1).

**Table (2-1) the host information of patients and control subject studied**

| Group | Case | No of cases | Female | Male | Age (Year) |
|---|---|---|---|---|---|
| I | Leukemia | 152 | 61 | 91 | 13-65 |
| a | ALL | 79 | 34 | 45 | 13-26 |
| b | CLL | 21 | 15 | 6 | 50-65 |
| c | AML | 7 | 5 | 2 | 14-30 |
| d | CML | 45 | 20 | 25 | 28-50 |
| II | Hodgkin's | 26 | 12 | 14 | 26-55 |
| III | Normal | 12 | 5 | 7 | 35-60 |
| IV | Nonmalignant | 37 | 15 | 22 | 20-60 |

ALL      Acute Lymphocytic Leukemia

AML      Acute Myelocytic Leukemia

CLL      Chronic Lymphocytic Leukemia

CML      Chronic Myelocytic Leukemia

## 2.4 Sera specimens:

Blood samples were collected from the patients and controls donors by venipuncture. The whole blood was left for 20 minutes at room temperature. After coagulation, the serum was separated by centrifugation at 3000 r.p.m for 10 minutes, serum specimens were frozen at- 20C° until assayed.

## 2.5 Determination of Biochemical constituents in sera of patients with leukemia, Hodgkin's and control non malignant.

### 2.5.1 Determination of total protein (TP) in sera of parients with Leukemia Hodgkin's and non malignant (control):

Protein content of the serum was determined by lowry method, using bovine serum albumin (BSA) as a standard protein[188].

**Reagents:**

1- Reagent A : Alkaline sodium carbonate solution, 2 gm of sodium carbonate ($Na_2 CO_3$) was dissolved in 100 ml of 0.1 N sodium hydroxide (NaOH).

2- Reagent B: copper sulphate – Na,k- tartrate solution 0.5 gm of copper sulphate (Cu $SO_4$. $5H_2O$) was dissolved in 100 ml of deionized water. From this solution, 10 ml was taken to be added to 0.1 gm of Na, k-tartrate, this reagent was prepared freshly in the day of use.

3- Reagent C: Alkaline copper solution: this reagent is prepared on the day of use by mixing 50 ml of reagent A and 1ml reagent B.

4- Reagent D: Follin – cio calteau reagent: This reagent is a solution of sodium tungstate and sodium molbydate in phosphoric acid and hydrochloric acid.

The reagent was prepared by dilution of the commercial reagent with an equal amount of deionized water on the day of use.

5- Standard protein: The standard protein (BSA), was prepared as follow:

   a- A stock solution of 1000 µg/ml, was prepared by dissolving 100mg BSA in 100ml deionized water.

b- From the stock solution , the following concentrations were prepared by serial dilutions with deionized water: 25, 50, 100, 150, 200 µg/ml.

## Procedure:

1- Two and half ml of reagent C was added to 0.5 ml of standard protein or diluted serum sample, and left for 10 minutes after mixing them throughly.

2- Tow hundred and fifty ml of reagent D was added and mixed immediately and rapidly. This mixture was left for 30 minutes, then the absorbance was read at 600 nm.

## Calculations:

The standard curve was obtained by plotting the absorbance against the corresponding concentration of the standard protein, and was used to determine the unknown protein concentration of the serum sample.

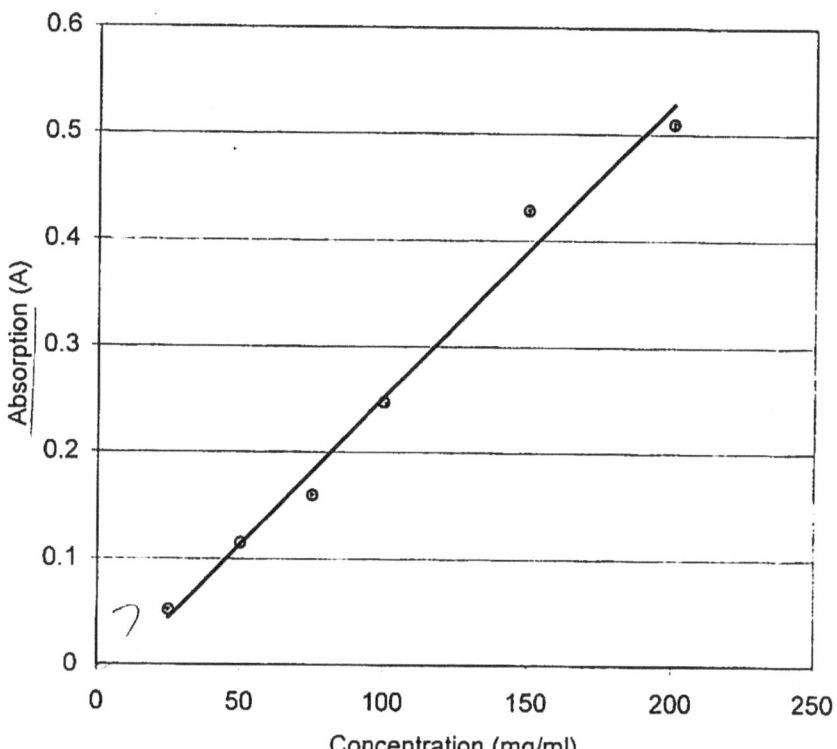

Figure (2-1)
The standard curve for determination of total
protein concentration.
All other details are explained in text.

**2.5.2 Determination of total sialic (TSA) Levels in sera of patients with leukemia, Hodgkin's and non malignant diseases (control):**

**Reagent:**

1- Resorcinol stock solution (2% w/v): 1 mg of resorcinol was dissolved in 50ml deionized water. The solution was prepared in the day of use.

2- Copper sulphate solution (0.1M): 4 gm of $CuSO_4$. $5H_2O$ was dissolved in 250ml deionized water.

3- Resorcinol reagent: This reagent was prepared in the day of use by mixing 5ml of resorcind stock solution, 4.875 ml deinized water, and 0.125 ml (0.1M) $CuSO_4$, $5H_2O$ solution, the final volume was made up to 50 ml with HCL (concentrated).

4- Butyl acetate/ methanol reagent (85: 15 v/v): 85 ml of butyl acetate was mixed with 15ml of methanol. This solution was stored in cold place.

5- Standard sialic acid: The standard sialic acid solutions with different concentration (5, 10, 15, 20, 25, 30, 35, 40, 45 µg/ml) were prepared by serial dilution from stock standard solution of sialic acid.

**Procedure:**

1- Twenty µl of standard sialic acid solutions or serum sample added to 980 µl of deionized water, the assay tubes was vortexed and placed on ice.

2- To each assay tube, 1ml of resorcinol reagent was added, then the tubes were placed in a boiling water for exactly 15 minutes, then for 10 minutes on ice.

3- Two-ml of butyl acetate/ methanol reagent was added, the assay mixtures were vortexed and centrifuged for 10 minutes at 3000 r.p.m. The extracted chromphore was read at 580nm, against deionized water.

**Calculation:**

The standard curve was obtained by plotting the absorbance at 580 nm. against the corresponding concentrations of standard sialic acid solutions, and was used to determine the TSA levels in the serum sample.[189]

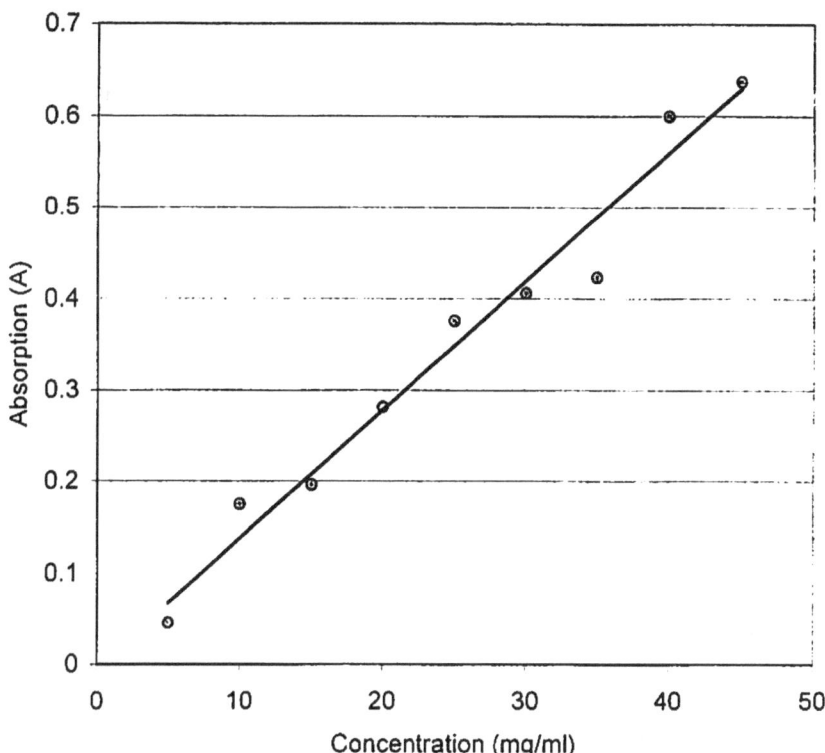

Figure (2-2)
The standard curve for determination of  total sialic acid
concentration.
All other details are explained in text.

45

*Chapter Two* ─────────────────── *Total and lipid - associated sialic acids levels in sera of Leukemic and Hodgkin's patients*

**2.5.3    Determination of lipid- associated sialic acid (LASA) levels in sera of Leukemic, Hodgkin's, and non malignant (control) patients:**

**Reagents:**

1- Chloroform/ methanol: this solution was prepared by mixing two volume of chloroform with one volume of methanol, it was store at 4 C°.

2- Phospho tungstic acid solution: 1 gm of phospho tungstic acid was dissolved in 1 ml of deionized water, then heated to produce clear solution, the solution was prepared on the day of use.

3- Resorcinol reagent and butyl acetate / methanol reagent were prepared as described in section (2.5.2) .

**Procedure:**

LASA was assayed in serum samples by using the method described by katapodis , et al.[189]. LASA assay consist of the following steps:

1- One hundred and fifty μl of deionized water added to 50 ml of serum sample, the tube was vortexted for five seconds and placed on ice.

2- Three ml of cold (4C°) chloroform/ methanol solution was added, after mixing them, 0.5ml of cold deionized water was added, the tubes were centrifuged for five minutes at 3000 r.p.m. at room temperature.

3- One ml of the resulting upper layer was transferred to another tube, and 50 μl of phospho tungestic acid solution were added, mixed and allowed to stand at room temperature for 5 minutes.

4- The assay tubes were centrifuged for 5 minutes at 3000 r.p.m, the supernatant fluid was removed and remaining precipitate was dissolved in one ml of deionized water. Then, the step 2 and 3 mentioned in section (2.5.2) were followed.

**Calculations:**

LASA levels in sera samples were determined using the standard sialic acid curve figure (2-2).

## 2.6 Statistical methods:

The results for TSA, LASA, TSA/TP ratio were analyzed statistically by values expressed as mean ± SD The level of significance was determined by student's- test[190].

**Results and Discussion:**

Serum samples from persons were tested for total protein (TP) total sialic acid (TSA), and lipid associated sialic acid (LASA) these included 178 patients with blood diseases (Leukemia, Hodgkin's) and 37 patients with non malignant disease and 12 normal healthy.

Table (2-2) shows the mean values ± SD of the four biomakers TP, TSA, LASA, TSA/TP obtained from sera of the controls and the groups of patients (Leukemia, Hodgkin's).

The results in table (2-2), show that the mean values of total protein (TP) in sera of Leukemic and other blood diseases patients did not find any significant differences, when compared to the mean value of normal and pathological controls and usually, the mean values of total protein in sera of all group still in normal range, but the levels of TP were generally increased in Leukemia and Hodgkin's in comparison with healthy group. The elevation of total protein levels in leukemia and Hodgkin's diseases are due to the destruction of blast cells[191].

Table (2-2): Sera total proteins, total sialic acid, lipid-associated sialic acid and sialic acid normalized to total proteins in leukemia, Hodgkin's patients, pathological control and normal controls

| Group | Samples No. | Total proteins (TP) gm/dL ± SD | Total sialic acid (TSA) mg/dl ± SD | Lipid associated sialic acid (LASA) mg/dL ± SD | Total sialic acid:total proteins TSA/TP mg/dL ± SD |
|---|---|---|---|---|---|
| Leukemia | | | | | |
| ALL | 79 | 7.33 ± 0.93 | 97.1±12.15 | 37.17±14.22 | 13.14± 4.3 |
| CML | 45 | 7.54 ± 0.46 | 91.67 ± 16.3 | 35.02 ± 14.2 | 11.64 ± 3.2 |
| CLL | 21 | 7.36± 1.14 | 85.28 ± 11.48 | 34.61 ± 12.6 | 11.21 ±3.53 |
| AML | 7 | 7.61 ± 0.28 | 91.42 ± 14.35 | 41.42 ± 10.69 | 13.8 ± 4.5 |
| Hodgkin's | 26 | 7.22 ± 0.53 | 80.65 ± 11.45 | 32.23 ± 10.27 | 11.11 ±3.25 |
| Pathological controls | 37 | 7.30 ± 0.79 | 59.86 ± 10.8 | 24.60 ± 6.72 | 8.81 ± 0.47 |
| Normal controls | 12 | 7.09 ± 0.34 | 50.30 ± 8.02 | 20.66 ± 3.82 | 7.14 ± 0.48 |

ALL   Acute Lymphocytic Leukemia
AML   Acute Myelocytic Leukemia
CML   Chronic Myelocytic Leukemia
CLL   Chronic Lymphocytic Leukemia

48

*Chapter Two* ――――――――――――――― *Total and lipid - associated sialic acids levels in*
*sera of Leukemic and Hodgkin's patients*

The results of table (2-2) also reveal that the mean values of sera TSA in leukemia and Hodgkin's patients were elevated ($P<0.001$ and $P<0.01$) in comparison to normal healthy and pathological controls ($P<0.001$ and $P<0.001$) respectively). However the comparison of sera TSA for different types of leukemia (ALL, AML, CLL, CML), the leukemia and Hodgkin's patients show high level of TSA as compared with pathological controls and normal healthy. There was highly significant ($P<0.001$) difference between the serum TSA in (leukemia, Hodgkin's) patients and normal healthy and even control groups, and no significant difference in sera TSA level in the four types of leukemia and Hodgkin's are observed. It is clear from table (2-2) that the mean values of total sialic acid to total protein ratio (TSA/TP) in sera leukemia and Hodgkin patients were significantly elevated ($P<0.01$) as compared to the normal healthy and control as shown in table (2-2). Accordingly, the TSA/TP ratio could be useful markers for detecting malignancies. Some authors have demonstrated a significant rise in TSA/TP and other biomarkers levels in cancer patients[192]. Cell death any where in the body protein induce local cellular and vascular reactions which might cause changes in the composition of extra cellular fluid[193][194].

Table (2-3) shows, the specificity and sensitivity of TSA test in normal individuals and patients with leukemia, Hodgkin's and control. The specificity was calculated as the number of cases having TSA level less or equal the upper limit of normal (73 mg/dL) divided by the total number of cases.

The sensitivity of TSA test was calculated by dividing the total number of cases which have TSA levels upper than the normal limit (73 mg/dL) divided by the total number of cases. The results from table (2-3) show that the more than 85% all kind of leukemic patients had elevated

49

*Chapter Two* ———————————— *Total and lipid - associated sialic acids levels in sera of Leukemic and Hodgkin's patients*

values of TSA, also Hodgkin's had nearly the same elevated values, on the other, hand the normal and the control group the specificity test of TSA was 59.4% and the normal group %100.

**Table (2-3) specificity and sensitivity of total sialic acid measurment**

| | Group | No of cases | Specificity % (*) | Sensitivity (**) |
|---|---|---|---|---|
| I | ALL | 79 | 12.65 | 87.34 |
| | CML | 45 | 11.1 | 88.8 |
| | CLL | 21 | 9.5 | 90.4 |
| | AML | 7 | 14.4 | 85.5 |
| II | Hodgkin's | 26 | 15.3 | 84.6 |
| III | Normal | 12 | 100 | -- |
| IV | Control | 37 | 59.45 | 40.54 |

(*) The number of cases have TSA values ≤ 73 mg/dL divide by the total number of cases by 100.

(**) The number of cases have TSA values > 73 mg/dL divided by the total number of cases by 100.

   ALL     Acute Lymphocytic Leukemia

   AML     Acute Myelocytic Leukemia

   CLL     Chronic Lymphocytic Leukemia

   CML     Chronic Myelocytic Leukemia

Figure (2-3) show the distribution of the individual values of TSA in sera of leukemic patients, Hodgkin's, normal and control

48

50

*Chapter Two* ———————————— *Total and lipid - associated sialic acids levels in sera of Leukemic and Hodgkin's patients*

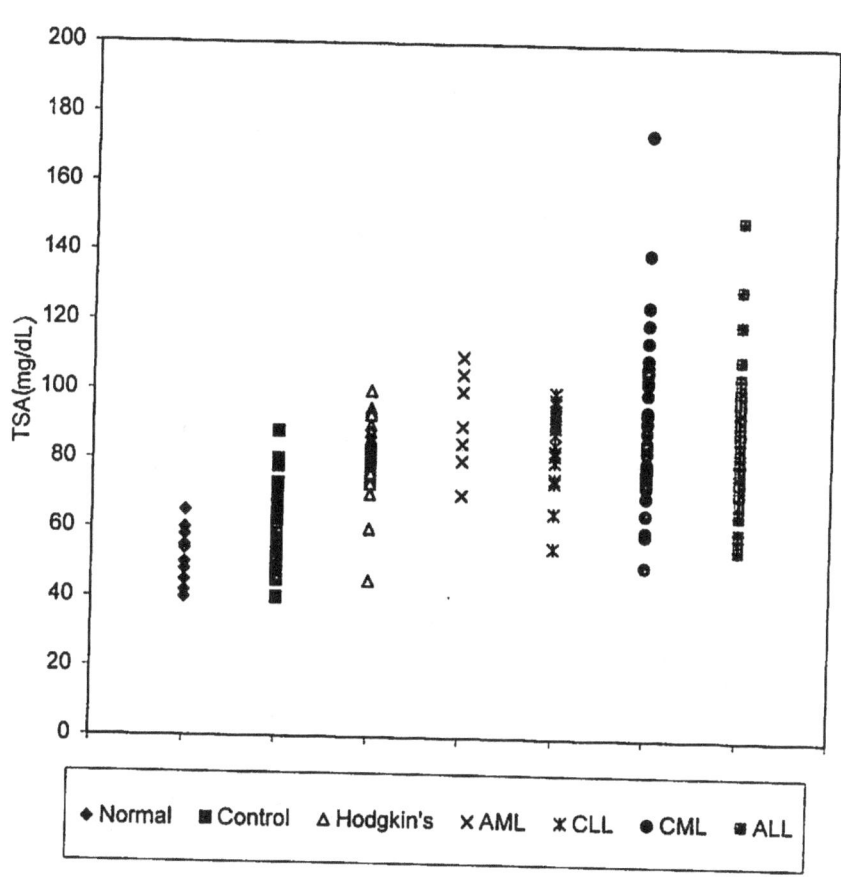

Figure (2-3 a)
Distribution of the individual values of TSA in sera of
leukemic  patients, Hodgkin's, control and normal healthy.
All other details are explained in text.

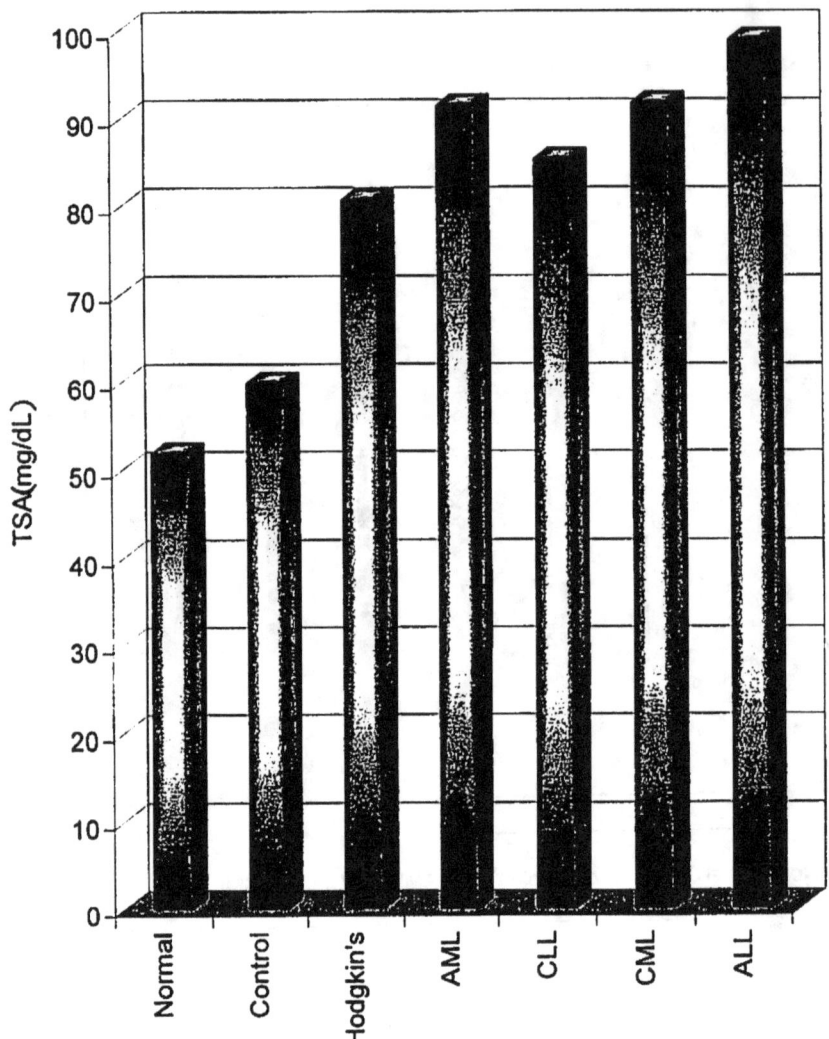

Figure (2-3 b)
The mean of the individual values of TSA in sera of
leukemic  patients, Hodgkin's, control and normal healthy.
All other details are explained in text.

52

*Chapter Two* ————————————— *Total and lipid - associated sialic acids levels in sera of Leukemic and Hodgkin's patients*

Many glycoproteins and glycolipids from malignant cells differ in carbohydrate composition from those found in normal cells since many of these glycoconjugates contain terminal sialic acid which can be shed into the circulation[86].

Cell surface glycoproteins and glycolipds have altered cell to cell recognition. Cell adhesion, antigenicity, and invasiveness in malignant conditions[95][196].

Increased sialic acid levels on the surface of transformed or malignant cell have been shown to be released into the circulation[196]. The potential role of sialic acid in the mechanism of tumor formation is indicated by the finding that sialic acid mask the surface of certain tumor cells by interfering with the immune response of the host[197] and that sialic acid content appears to be correlated with metastatic ability in variety tumor cells[198].

Total sialic acid (TSA) is of great interest as marker of malignancy although it has been demonstrated to be specific for any type of cancer[49]. Serum TSA levels have been found to be elevated in a number of different cancers[50][52][55]. Studies by Shamberger and Dnistrain on adults affected with neoplasms of the hemopoietic system, point the usefulness of determining sialic acid (SA) in monitoring the neoplastic process[64][112].

Increased concentration of sialic acid were noted in neoplasmas of the lungs[179][186] gall bladder, melanoma[79], and neoplasm of the female sex organs[199].

Many authors analyzing the concentration of sialic acid (TSA) in various neoplasms stress the fact that reaches higher levels of (TSA) (higher concentration of sialic acid), with the advance of the neoplasm and increases during metastasis[186].

Normalization of silaic acid concentration after treatment of the disease constitutes a favorable prognosis while it s increase creates suspicion of relapse of the disease[186][200][201].

In patientsaffected by ovarian cancer it was noted that concentrations exceeding (120 mg/dL) were related to poor prognosis[202]. Comparison the results in (Figure. 2-4) with staging of the leukemia (ALL, CML) diseases showed a good agreement between sialic acid concentration (TSA) and the course of disease[202].

Increased (TSA) levels (sialic acid concentration) after the end of treatment in comparison with the control group may be a result of minimal residual disease[203]. In cases of relapse we observed again an increase of sialic acid concentration. It value remained high if treatment failed[203].

Lack of decrease in sialic acid concentration constituted the basic for poor prognosis, figure (2-4) show that TSA elevated to high in last stage before death, for ALL and CML patients 3 patients with (ALL) and 1 with CML, a serial determination for TSA was made one of the ALL patient who died and the CML patient also who died with disease, had a significantly elevated in TSA. This gives us a good tool for prognosis and monitoring stager of disease and the response for treatment.

Because sialic acid isaspecific marker for a given type of neoplasm changes in concentration can only be interpreted together with the clinical picture of the diseases.

54

*Chapter Two* ———————————— *Total and lipid - associated sialic acids levels in sera of Leukemic and Hodgkin's patients*

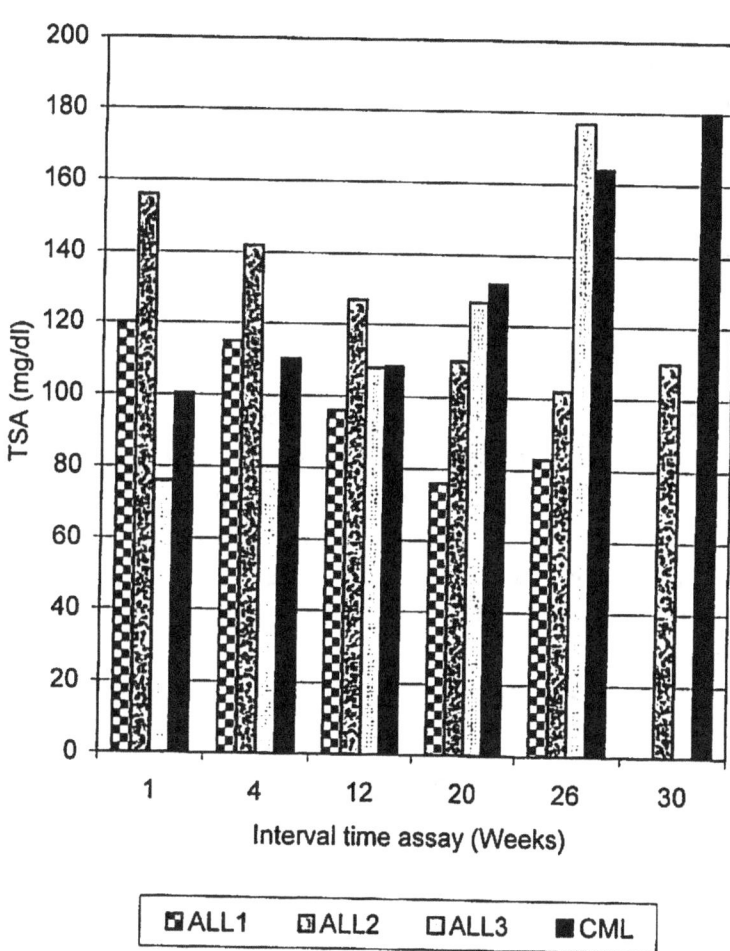

Figure (2-4)
TSA concentration in different stages of leukemic
patients
All other details are explained in text

## Determination of lipid-associated sialic acid (LASA) levels in sera of patients with leukemia, Hodgkin's and non-cancer individuals:

Serum lipid- associated sialic acid (LASA) levels were determined in normal healthy persons and in non-cancer (control) patients, using the method of katopodis (et al),[190].

Table (2-2) shows the mean values of LASA in the four types of leukemia, Hodgkin's control and normal individuals, and the comparison of the mean values of sera LASA between these groups. The result in this table reveal over all elevation in LASA levels for each group of cancer compared with normal healthy and control individuals.

The increased LASA levels in sera with leukemia, Hodgkin's were statistically significant when compared with those obtained from normal healthy and control patients (0.001, 0.001) respectively.

Table (2-4) represents the percentages of LASA specificity and sensitivity in leukemia, Hodgkin and normal healthy individuals and control patient, calculated by the value of 19 mg/dL as the upper limit of normal.

In general. high sensitivity was observed 100% for all types of leukemia and 96.1% for Hodgkin's 72.9% for control (non-malignant) and for normal healthy 33.3%.

**Table (2-4) Specificity and sensitivity of LASA in leukemia, Hodgkin's normal, and control (malignant) patients**

| | Group | No of cases | Specificity % (*) | Sensitivity (**) |
|---|---|---|---|---|
| I | ALL | 79 | --- | 100 |
| | CML | 45 | --- | 100 |
| | CLL | 21 | --- | 100 |
| | AML | 7 | --- | 100 |
| II | Hodgkin's | 26 | 2.85 | 96.15 |
| III | Non malignant (control) | 37 | 17.1 | 72.9 |
| IV | Normal healthy | 12 | 66.6 | 33.3 |

(*) Specificity was calculated by division the number of cases having LASA value 19.84 mg/dL by the total number of cases by 100.

(**) Sensitivity was calculated by of division the number of cases having LASA value > 19.84 mg/dL by the total number of cases by 100.

ALL    Acute Lymphocytic Leukemia

AML    Acute Myelocytic Leukemia

CLL    Chronic Lymphocytic Leukemia

CML    Chronic Myelocytic Leukemia

56

*Chapter Two* ———————————— *Total and lipid - associated sialic acids levels in sera of Leukemic and Hodgkin's patients*

Thus, serum LASA has been suggested as useful tumor marker, more satisfactory than total sialic acid[177][201].

Lipid associated sialic acid LASA is expected to behave as a non-specific tumor marker after shedding, and that is an important aspect of the turnover of normal cell surface constituents which occurs principally in both normal and malignant growing cell[203].

The present investigation showed significantly elevated levels of LASA in sera of cancer patients (Leukemia, Hodgkin's), when compared with control, and normal healthy individual. This finding is in agreement with other studies which have reported an elevation of serum LASA in cancer patients[203][204]. Figure (2-5a) shows the distribution of individual values of LASA in sera for Leukemia, Hodgkin's and control. Dwivedi, et al.[60], measured the LASA levels in eight different cancers (lung, colon, ovarian, prostate, Leukemia thyroid, pancreas, adrenal and gastrointestinal cancer), the results were compared with normal healthy. It shows that LASA is a useful marker, as a prognostic determinant in a variety of neoplastic condition.

A serial determination of LASA was made for 3 colon 1 prostate 1 ovarian cancer patients, the two patients with colon cancer who died with the metastatic disease had significantly elevated LASA values and the prostate cancer also died with the ovarian patients plasma, LASA was decreased after surgery[60]. .

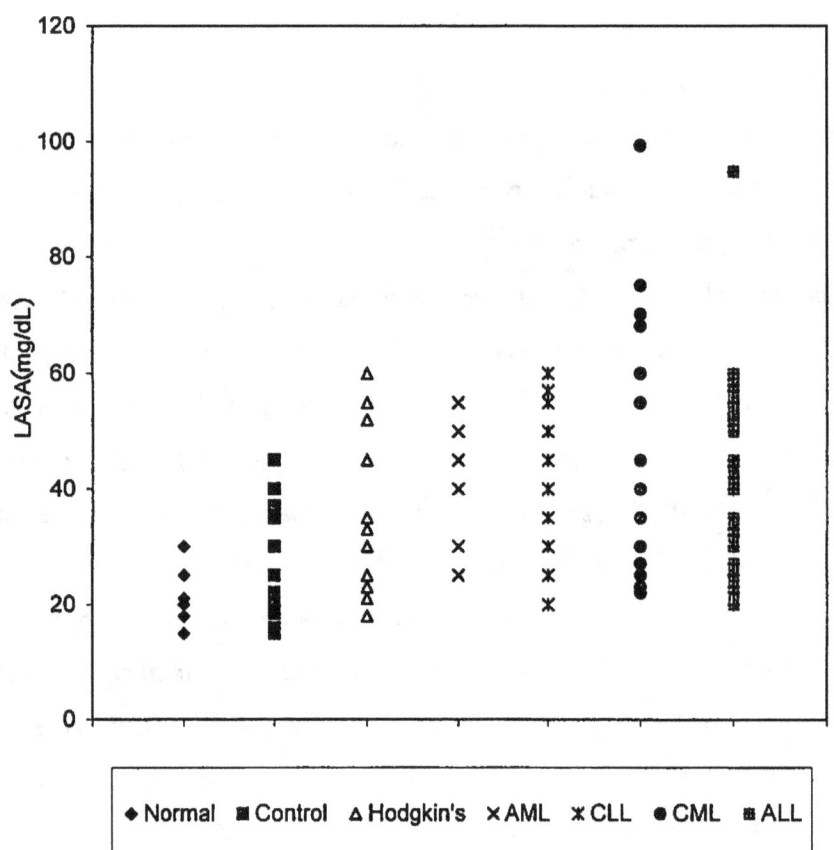

Figure (2-5 a)
Distribution of the individual values of LASA in sera of
leukemic patients, Hodgkin's, control and normal healthy.
All other details are explained in text.

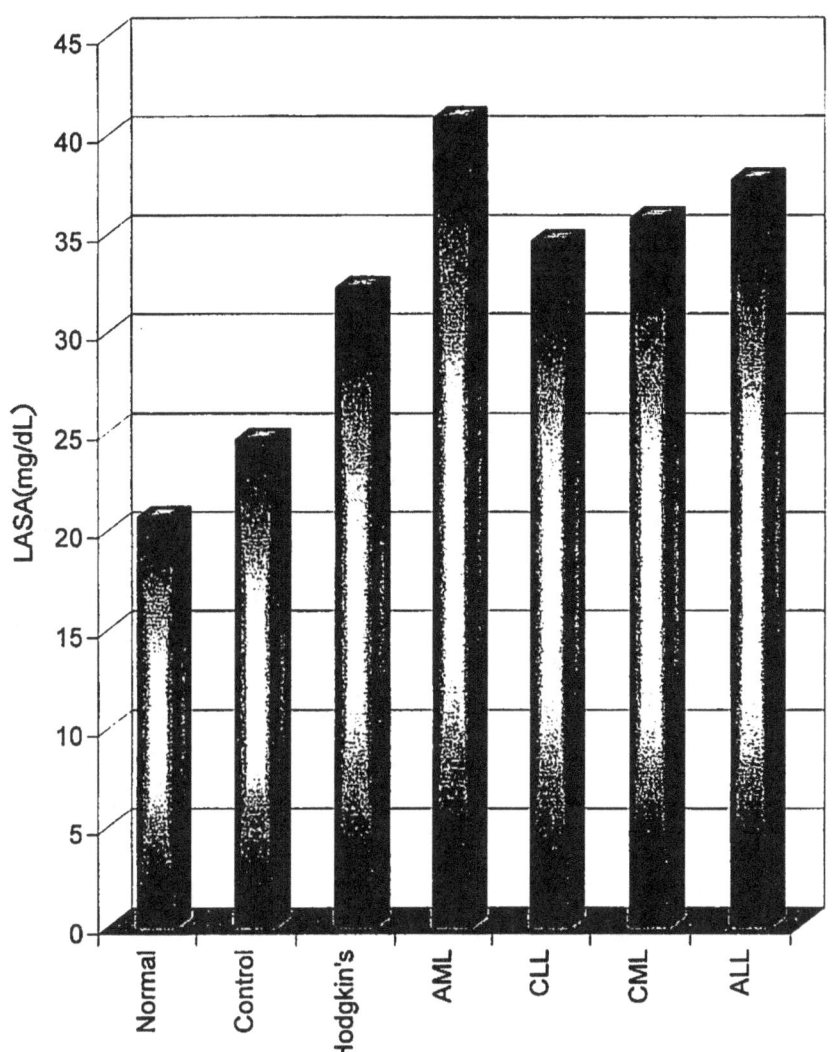

Figure (2-5 b)
The mean of the individual values of LASA in sera of
leukemic  patients, Hodgkin's, control and normal healthy.
All other details are explained in text.

Elevated LASA levels were observed with progression of the disease. The LASA level was increased in more patients with Leukemia, and Hodgkin's, compared with control and normal healthy. In a recent study[187] it has been reported that there is a variation in LASA serum levels related to disease extent i.e the more advanced the disease, the higher is the LASA serum value, in the following order local < loco regional < metastatic disease. This finding can be taken as a proof that LASA can be as a marker for the assessment of disease extent.

In another studies, LASA levels have been reported to be a useful in monitoring patients with malignant melanoma[205]. Elevated LASA levels in breast cyst fluid have been associated with increased risk of breast cancer[183][206]. Some conditions, other than cancer, affect LASA. Among them, are myocardial infarction[207], infection, rheumatoid arthritis, and collagen degeneration. Polivkova et al.[73], believe that the determination of LASA levels could be useful not only for cancer diagnosis but also prognosis.

Toumbis et al.,[208] did not find any significant differences in LASA levels between benign and malignant pleural effusions.

Some authors used, TSA and TSA/TP and LASA in diagnosis of children with various malignancies, a significantly elevated TSA and TSA/TP and LASA were observed in different tumors, a decreasing of TSA, TSA/TP and LASA in neurablastoma and Yolk sac tumors after a successful treatment of the malignancy[80].

From our results, LASA appears to be useful marker in distinguishing between healthy individuals and cancer patients, since LASA exhibits a high sensitivity. The combined use of LASA with other markers may provide high degree of marker positively.

Figure (2-6) shows a serial determination of LASA that was made in four Leukemic patients, 3 ALL and 1 CML, the data show that one of the ALL, patient and the CML patient who died with disease had a significantly elevated LASA values, and on the other hand the LASA had a slightly decreased in patients who were responding to chemotherapy, the LASA levels of Leukemia is > Hodgkin.

60

*Chapter Two* ————————————— *Total and lipid - associated sialic acids levels in sera of Leukemic and Hodgkin's patients*

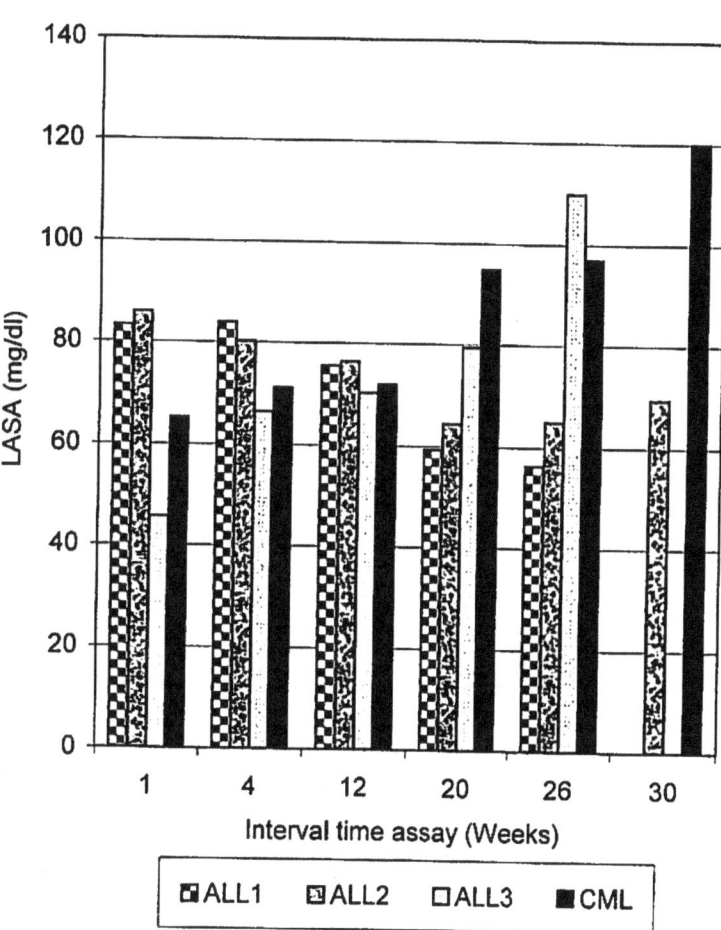

Figure (2-6)
LASA concentration in different stages of leukemic
patients
All other details are explained in text

# CHAPTER THREE 3

## Glycoproteins levels in sera of Leukemic and Hodgkin's patients

Glycoproteins and glycolipids are useful indicators in diagnosis and management of cancer[51] [102].

There is ample evidence to show that modifications in glycosylation of cell surface glycoproteins and glycolipids are often associated with malignant transformation of cell[180]-[182]. Variations in glycoprotein moieties like fucose, hexosamines and seromucoid fraction are found to be significantly useful as a tumor marker in different neoplastic disease[170][209][210].

Previous studies have examined derived glycoprotein markers, including bound sialic acid, seromucoid fraction, carcino-embryonic antigen and alpha-1acid glycoprotein in patients receiving anticancer treatment and found that the markers have potentials to be used as substitutes for the existing criteria for retrospective judgement of response to treatment[211][213].

Thus the glycoprotein levels, as indicated by sialic acid fucose and seromucoid fraction, are useful serological biomarkers with regard to diagnosis, classification, staging, prognostication and treatment monitoring of patients with malignant disease. Receptors on these glycoprotein constituents in Leukemia patients are very few like any other malignancy[66], early detection helps in better management in Leukemia patients. However, problems in early diagnosis and treatment monitoring of Leukemia patients arise due to the lack of parameters sensitive enough and specific for Leukemia conditions[66].

Glycoproteins play an important role in the cellular phenomena that undergo alterations during cancerous transformation[214] Patel. et al. demonstrated that seromucoid and fucose could be useful for early diagnosis, monitoring of Leukemia[66]. However, the protein-bound carbohydrate and seromucoid of plasma have been demonstrated to be elevated in wide variety of pathological conditions including spontaneous human carcinoma[196].

The present investigation was carried out to clarify further the possible usefulness of serum protein-bound sugars and seromucoid as for possible identification of the Leukemia and other blood cancer.

# Materials and Methods

## 3.1 Chemicals:

All chemicals and reagent mentioned in <u>section (2-1)</u> of chapter 2 were used in the experiments of this chapter.

## 3.2 Instruments:

All instruments that described in <u>section (2-2)</u> of chapter 2 were used in the experiments of this chapter.

## 3.3 Patients and blood samples:

Twenty-two patients with ALL and 13 with CML, 10 with CLL, 5 with AML, 10 with Hodgkin's, 11 with non-malignant disease (control) and 8 normal healthy, were used in this study.

## 3.4 Preparation of stock solutions:

The stock solutions, that used in experiments of this chapter are:

1- Perchloric acid solution (1.8M): 16.6ml of 72% perchloric acid was diluted to 100ml with distilled water.

2- Phosphotungstic acid solution, 5% in 2N of HCl.

3- Sodium chloride solution, 0.85% (0.85gm of NaCl was dissolved in 100ml of distilled water).

4- Sodium hydroxide solution, 0.1N (0.4gm of NaOH was dissolved in 100ml of distilled water).

5- $H_2SO_4$, 60% (V/V): 60ml of $H_2SO_4$ was mixed with 40ml of distilled water.

6- Orcinol reagent (2%): 2gm of recrystallized orcinol was dissolved in 100ml of 30% (V/V) $H_2SO_4$.

7- Stock standard sugar: 100mg of galactose and 100mg of mannose were dissolved in 100ml of distilled water, then saturated with benzoic acid and stored in cold place.

8- Working standard: 1ml of stock standard sugars was mixed with 9ml of distilled water, this reagent was prepared freshly in the day of use.

## 3.5 Determination of glycoproteins in sera of patients with Leukemia, Hodgkin's, control and normal healthy individuals.

### 3.5.1 Determination of Seromucoid:

The method of Weimer and Mashin[215] was used to determine serum seromucoid. This method includes the following steps:

1- Half ml of serum was added to 4.5ml of 0.85% NaCl, then mixed and 2.5ml of 1.8M perchloric acid was added after mixing by inversion, the assay tubes were allowed to stand at room temperature for 10 minutes, then centrifuged for 15 minutes at 3500 r.p.m to obtain clear supernatant.

2- To 5ml of the supernatant, 1ml of phosphotungstic acid reagent was added, after mixing the tubes were allowed to stand for 10 minutes.

3- The tubes were centrifuged after removing of the supernatant, 5ml of 95% ethanol was added, centrifuged and the supernatant was removed.

4- The resulting precipitate was dissolved in 0.5ml of (0.1N) NaOH, this was considered as the unknown.

5- Set up blank with 0.5ml of distilled water, and standard using 0.5ml of working standard.

6- To each unknown, blank and standard, added 1.25ml of orcinol reagent and 7.5ml of (60% V/V) $H_2SO_4$.

7- All tubes were placed in a water bath at $(80 \pm 0.5\overset{\bullet}{C})$ for 20 minutes, cooled and read against distilled water at 520nm.

## Calculations:

$$\text{mg seromucoid/100ml} = \frac{Ax - Ab}{As - Ab} \times 0.1 \times \frac{100}{0.333} = \frac{Ax - Ab}{As - Ab} \times 30$$

Where

Ax: The absorbance of unknown solution at 520nm.

As: The absorbance of standard solution at 520nm.

Ab: The absorbance of blank solution at 520nm.

### 3.5.2 Estimation of serum protein-bound hexose:

The method includes the following steps:

1- One tenth of ml of serum was added to 5ml of 95% V/V ethanol and mixed carefully, then centrifuged for 15 minutes at 3500 r.p.m, after that the supernatant was decanted.

2- The remaing precipitate was washed with 5ml of 95% ethanol then stired, after that centrifugation.

3- The supernatant was decanted, the steps 4-7 was carried out as in section $(3.5.1)^{[215]}$.

## Calculations:

$$\text{mg protein bound hexose/100ml} = \frac{Ax - Ab}{As - Ab} \times 0.1 \times \frac{100}{0.1} = \frac{Ax - Ab}{As - Ab} \times 100$$

Where:

Ax = The absorbance of unknown solution at 520 nm.

As = The absorbance of standard solution at 520 nm.

Ab = The absorbance of blank solution at 520 nm.

## Results and discussion:

The levels of mucoid proteins and protein- bound hexoses were determined in sera of leukemia, Hodgkin, a non malignant (control) using the method of Weimer and Mashin[215].

The mean concentrations of proteins and protein bound hexoses in sera of all patients and normal healthy controls are summarized in table (3-1) and (3-2) Figures (3-1) and (3-2) show the distribution of the individual values of mucoid proteins and protein- bound hexose respectively.

From these results, statistically significant elevations were observed in the serum levels of mucoid protein of four different types of leukemia, and Hodgkin's, compared to normal healthy and even control patients ($P<0.001$, $P<0.01$) and ($P<0.01$, $P<0.1$) respectively. The mean concentration of mucoid proteins (MP) reached to 20.5 mg/dl in ALL and 22.66 mg/dl in CML and 21.3 mg/dl in CLL and 21.66 mg/dl in AML and 20.5 mg/dl in Hodgkin. Only minor differences in (MP) concentrations were observed when the four groups of leukemia were compared to each other: Table (3-3) shows the specificity and sensitivity of the seromucoid (MP). Test specificity and sensitivity were calculated considering the value 11.46 mg/dl as the upper limit of normal. This test was sensitive for different types of leukemia and Hodgkin's as compared to the normal healthy individuals. Test sensitivity in those with leukemia (ALL, CML, CLL, AML) were 90.9%, 84.6%, 80%, 80%) respectively and for Hodgkin 80%. Test specificity for different type of leukemia (ALL, CML, CLL, AML) were 9.09%, 15.3%, 20%, 20%) respectively and for Hodgkin 20%. From these results, it possible to conclude that it was useful for differential diagnosis and monitoring disease.

Table (3-1) Test specificity and sensitivity of sera levels of mucoid

proteins (MP) in patients with leukemia, Hodgkin, control,

and normal healthy.

| Group | | No of cases | Mean value ± SD mg/dL | Sensitivity % (**) | Specificity % (*) |
|---|---|---|---|---|---|
| I | ALL | 22 | 20.54±7.65 | 90.9 | 9.09 |
| | CML | 13 | 22.669 ± 7.97 | 84.6 | 15.3 |
| | CLL | 10 | 21.3 ± 7.52 | 80 | 20 |
| | AML | 5 | 21.06±8.107 | 80 | 20 |
| II | Hodgkin's | 10 | 20.55±9.35 | 80 | 20 |
| III | Control | 11 | 13.5±1.83 | 72.27 | 27.72 |
| IV | Normal | 8 | 11.763±3.35 | 37.4 | 62.5 |

(*) Calculated as the number of cases having MP ≤11.46 mg/dl divided by the total number of cases by 100.

(**) Calculated as the number of cases having MP > 11.46 mg/dl divided by the total number of cases by 100.

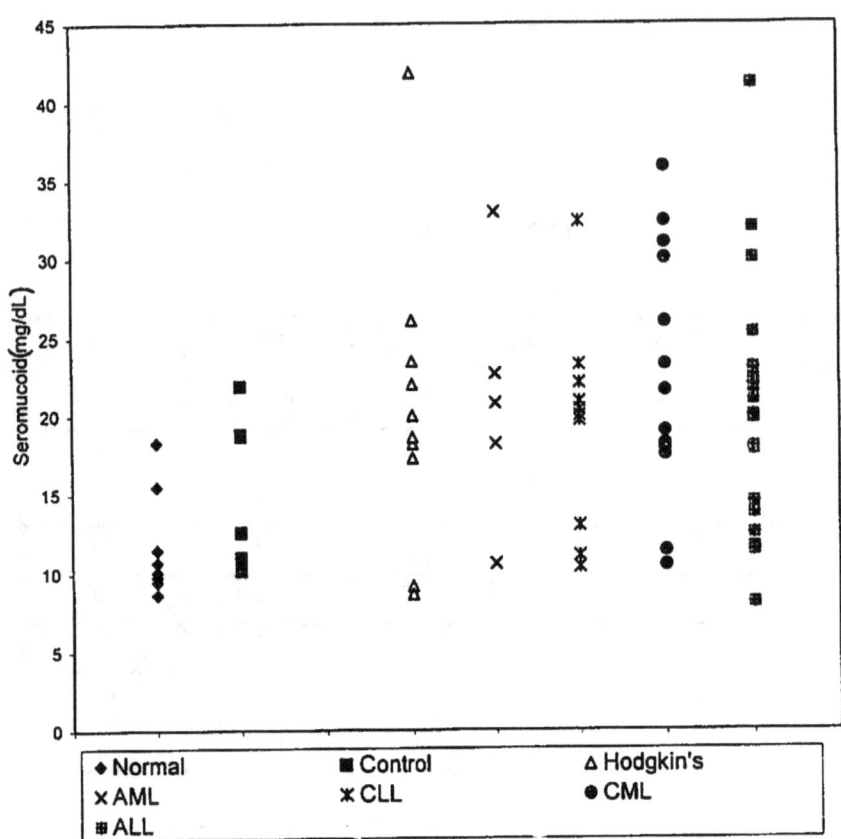

Figure (3-1)
Distribution of the individual values of Seromucoid in sera
of leukemic patients, Hodgkin's, control and normal healthy.
All other details are explained in text.

**Table (3-2) Test specificity and sensitivity of sera levels of protein-bound hexoses in patients with leukemia, Hodgkin, control, and normal healthy.**

| Group | | No of cases | Mean value ± SD mg/dL | Sensitivity % (**) | Specificity % (*) |
|---|---|---|---|---|---|
| I | ALL | 22 | 196.31 ± 17.5 | 86.3 | 13.6 |
| | CML | 13 | 211.82 ± 32.4 | 84.6 | 15.3 |
| | CLL | 10 | 192.7 ± 25.7 | 90 | 10 |
| | AML | 5 | 209.29 ± 40.2 | 80 | 20 |
| II | Hodgkin's | 10 | 182.04 ± 31.8 | 80 | 20 |
| III | Control | 11 | 118.66 ± 18.3 | 54.5 | 45.4 |
| IV | Normal | 8 | 92.725 ± 6.79 | 25 | 75 |

(*) Calculated as the numbers of cases have protein- bound hoxoses $\leq$ 118.01(mg/dl) divided by the total number of cases by 100.

(**) Calculated as the numbers of cases have protein- bound hoxoses > 118.01(mg/dl) divided by the total number of cases by 100.

Table (3-2) reveals that the mean values of protein –bound hexose in sera of patient of Leukemia, Hodgkin's were elevated significantly in comparison to those of normal and control patients.

The mean concentrations of protein- bound hexose reached 196.31 mg/dl in ALL patients, 211.82 mg/dl in CML patients, 192.70 mg/dl in CLL patient, 209.2 mg/dl in AML patients 182.04 mg/dl in Hodgkin's patients.

Table (3-2) shows the specificity and sensitivity of the protien- bound hexose in Leukemia, Hodgkin's, control and normal healthy. In the test using 118.01 mg/dl as the upper limit of normal, test sensitivity in those with Leukemia was varied from 80% to 90%, and 80% in Hodgkin's patients. The test was sensitive when compared to the normal healthy. Also, the results show low specificity in different types of Leukemia and Hodgkin's patients.

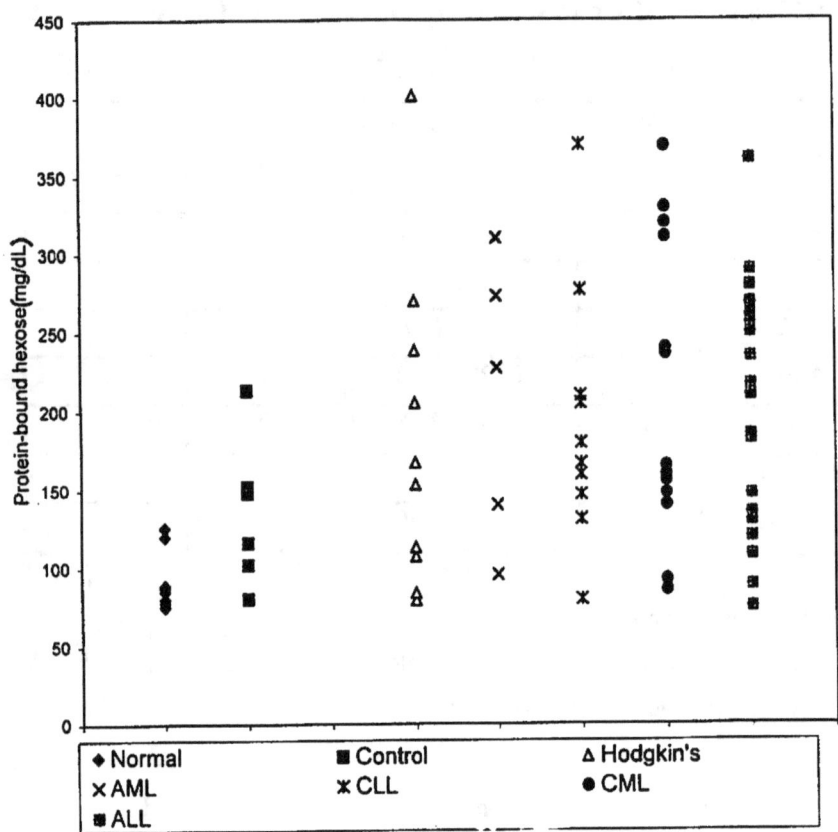

Figure (3-2)
Distribution of the individual values of Protein-bound
hexose in sera of leukemic patients, Hodgkin's, control and
normal healthy.
All other details are explained in text.

From these results, it is possible to conclude that sero mucoid protein was useful for differential diagnosis and monitoring of disease. Elevated level of glycoproteins have been reported in the sera of cancer patients by many investigators, to clarify the possible usefulness of seromucoid proteins (MP) and protein- bound hexose as a bio markers for identifying cancer diseases[140]. The mucoid proteins and carbohydrate-rich fraction of proteins have been studied as an indicator of tumor presence and the changes in glycoproteins could be determined by their carbohydrate moieties[140][216]. Variations in glycoproteins in human uterine hexosamine and sialic acid[217].

Patel, PS. et al.,[66] has observed significant increase in the levels of the protein- bound hexose and seromucoid in Leukemia and anemia patients when compared to the controls, it has suggested that the measurement of the two parameters could be helpful in the diagnosis and management of human Leukemia and anemia.

Yamanooto, et al[218] reported that the increase in protein- bound hexose arises as a result of depolymerization of the ground substance connective tissue adjacent to cancer with release of these compound into circulation.

Bolmer, et al.,[219] have suggested that the elevation of the plasma glycoprotein reflects merely the occurrence of tissue destruction , while Yashiko, et al[220] have concluded that tissue proliferation or repair is a more probable etiological factor. Since the results show a significant increase in the level of glycoprotein such as protein- bound hexose and seromucoid in sera of different type of Leukemia and Hodgkin's, it is suggested that the glycoprotein test may be used as a clinical marker for diagnosis of the disease.

An ideal tumor marker should not only have the potential for discriminating patients with malignancy from healthy individuals, but the alterations should be specific enough to differentiate between non- malignant and malignant conditions (pathological control and cancer patients) as well.

In the present study, the levels of the markers were also significantly higher in Leukemia, Hodgkin's patients compared to control (non- malignant) patients and normal healthy individuals.

# CHAPTER FOUR

## Determination of some biochemical constituents in sera of Leukemic and Hodgkin's patients

Lactate dehydrogenase (LDH) is an hydrogen transfer enzyme that catalyzes the oxidation of lactate to pyruvate with the mediation of $NAD^+$ as hydrogen acceptor. The reaction is freely reversible[221].

CH$_3$
|
H—C—OH  + NAD$^+$     $\xrightleftharpoons[\text{pH 7.4 - 7.8}]{\overset{\text{LDH}}{\underset{\text{pH 8.8 - 9.8}}{}}}$     CH$_3$
|
C=O  + NADH  + H$^+$
|
C=O
|
O$^-$

C—O
|
O$^-$

The (LDH), is one of the non-plasma specific enzyme. The concentration of this enzyme in tissues is very high, compared with those in sera, and its tissue level *is* about 500 times greater than those normally found in serum, and the leakage of the enzyme from even small mass of damage tissue can, increase the observed serum level of (LDH) to a significant extent[222][223].

The elevation of lacate dehydrogenase activity in malignancy is non-specific. It has been demonstrated in a variety of cancer, including Liver, non Hodgkin's lymphoma acute leukemia and other cancers[29]. Such as, breast, colon, stomach and lung cancer[224][231].

75

*Chapter Four* ——————————— *Determination of some biochemical constituents*
*in sera of Leukemic and Hodgkin's patients*

An increase in serum (LDH) activity occurs in about 2/3 of patients with leukemia and those with solid tumors, particularly wide spread or rapidly growing[232]. In Leukemia's, the serum (LDH) levels correlated with higher initial leukocyte counts but not with treatment outcome[233][236].

Recently (LDH) level is used but associated with other parameter in the monitoring the response to chemotherapy, irradiation therapy and bone marrow transplantation[237]-[250].

Different methods have been applied to the assay of LDH, such as spectrophotometric, coloriemtric and fluorometric, but the most frequently used method is the spectrophotometric[223]. This method depends on measuring the inter conversion of the NAD and NADH at 340 nm[251][252].

The reaction is either followed in the lactate to pyruvate (L$\longrightarrow$P) or pyruvate to lactate (P$\longrightarrow$L) direction.

The common methods which are routinely used for LDH assay are Wroblewiski, Ladeu, Wacker, Scandinavian and Henry method,[223][253]. These methods differ from each other in the direction of the reaction followed, type of buffer used, temperature and concentration of substrates. In this (Scandinavian) method, the reaction is followed by measuring the rate of NADH consumption at 340 nm[253].

Superoxide dismutase (SOD) is an enzyme that catalyze, the breakdown of ($\overset{\bullet}{O}_2$), and play important role in protecting the cells against damage by highly reactive superoxide free radical. SOD is found to be present in all oxygen metabolizing systems, and in most oxygen tolerant organisms[254].

It is particularly, likely to be, formed in the red cell and has been shown to be produced when oxyhemoglobin is autoxidized to methemoglobin[254][256]. Manganese superoxide dismutase (Mn SOD) also has been postulated as one possible mechanism of radio protection for hematopoietic cells[257].

It is also found a decrease in the activity of SOD in all malignant tumors and in sera of neoplastic diseases and juvenile chronic arthritis[204][258].

Oxidative damage may be involved in the pathogenesis of major disease such as cancer, atherosclerosis[259], and certain neurological disorders[260]. Inactivation and removal of reactive oxygen species depend on reactions involving the antioxidative defense system. The capacity is determined by a dynamic interaction between individual components, which include vitamins A,E and C and several antioxidative enzyme[261]. The most important enzymatic antioxidants are superoxide dismutase (SOD) which catalyzes dismutation of the super oxide anion $(\dot{O}_2)$ into $H_2O_2$, which is then deactivated to $H_2O$[261].

Super oxide is formed by the one- electron reduction of oxygen, and has been identified as a product in a number of biological reactions[262].

It is particularly likely to form in the red cell and has been shown to be produced when oxy hemoglobin is outoxidized to methemoglobin[263].

$$Hb - Fe^{+2} + O_2 \rightarrow Hb - Fe^{+3} + \dot{O}_2$$

Other likely sources include reactions initiated by ionizing radiation. Super oxide dismutase was used to show that the oxidation of epinephrine to adenochrome by xanthine oxidase is mediated by the super oxide radical[264]. Decreased activity of the enzyme superoxide dismutase (SOD) has also been found in all malignant tumors investigated so far[265]-[267]. Such changes in the contact of glycolipids (LASA) and the activity of (SOD) are not found in patient with benign tumors[268].

77

*Chapter Four* ——————————————— *Determination of some biochemical constituents in sera of Leukemic and Hodgkin's patients*

Recently SOD activity is used in monitoring the UV irradiation of treatment leukemia patients with cytotoxic drug[269][280].

The present investigation was carried out to compare the content of (LASA) and activity of SOD in sera of patients with leukemia, Hodgkin's and the normal healthy individual.

Trace elements are those elements that occur in human and animal tissue in milligram per kilogram or less[281]. Correspondingly, human intake requirements of trace are expressed in grams or fractions of grams per day, whereas human requirement of trace elements reported in milligrams per day. Since the 1970s, the term ultra trace elements has been used as microgram per kilogram or less and estimated dietary requirements indicated by micrograms per day[281].

Although, many details of trace elements function are not understood, some general characteristics are well known. These are amplification of trace element action, specificity, homeostasis, and interactions[183]. The trace elements occur primarily in combination with proteins and are also frequently considered separately[281].

The action of a very small amount of a trace element is necessary for optimal performance of the whole organism. Lack of a small amount of trace (e.g iron) can result in clinical abnormalities (anemia), seemingly disproportionate to the amount of element missing[281]. The basis for this amplification of trace element action are constituents, with, enzymes and hormones that regulate the metabolism of much larger amount of biochemical substance[281].

Any analytical method used for the determination of trace and ultra trace element in biological specimens must be sensitive specific, precise and relatively fast. Analytical sensitivity is very important because concentrations

**78**

*Chapter Four* ——————————— *Determination of some biochemical constituents*
*in sera of Leukemic and Hodgkin's patients*

of trace or ultra trace elements in some samples are in the non-organ per gram to microgram per gram range.

The most popular techniques for the determination of trace elements in biological specimens currently used include photometry, atomic absorption spectrophotometry (AAS), and emission spectroscopy (ES), and in our investigation we used the Atomic absorption[284][285].

Among the functions of the electrolytes are the maintenance of osmotic pressure and water distribution in the various body fluid compartments, maintenance of the proper pH, regulation of the proper function of the heart and other muscles, involvement in oxidation- reduction (electron transfer) reactions and participation in catalysis as cofactors for enzymes. Thus, it becomes quite apparent that abnormal levels of electrolytes and trace elements may be either the cause or the consequence of variety disorders and, that determination of electrolytes is one of the most important functions of the clinical Laboratory[286].

In leukemia $(K^+)$ elevated is due to the lysis of cell (hyperkalemia), while renal tubule toxins occasionally cause (hypokalemia) decrease $K^+$ levels[3].

Hypercalicimia is present occasionally and is due to bone destruction, it may be attributable to a parathomone - like material ectopically produced by the blast cell[5].

Accordingly, very limited amounts of work have been carried out on the role of trace elements, LDH, SOD and spectral studies in leukemia so, this part was carried out to study the effect of LDH, SOD, and trace elements then sera spectra studies in different types of leukemia and Hodgkin's disease.

# Materials and methods

## 4.1 Chemicals:

All chemicals and reagents mentioned in sectional (2.1) chapter 2 were used in the experiments of this chapter.

## 4.2 Instruments:

All instruments that described in section (2-2) chapter 2 were used in the experiments of this chapter.

## 4.3 Buffers and substrate:

1- Working tris buffer (0.056 M): 6.8 gm of tris (hydroxy methy) amino methane is dissolved in approximately 800ml of distilled water, the pH is adjusted to 7.4 with HCL and the volume of solution is brought to 1L by distilled water. This solution is stable for at least 6 weeks at 4C°.

2- Tris- NADH reagent: 13 mg of NADH is dissolved in 90ml of working tris- buffer. The absorbance is measured and brought to an absorbance of 1.0 (161 mM) by dilution with the same buffer, this solution is stable for 72 hours at 4C°.

3- Pyruvate working solution (13.5 mM): 149 mg of sodium pyruvate is dissolved in 100 ml of distilled water. This solution is stable for 20 days at 4C°.

4- Phosphate buffer (0.067 M), pH 7.8 .

5- EDTA solution (0.1M), containing 1.5mg KCN per 100 ml.

6- Riboflavin solution (0.12M) 4.5mg of riboflavin dissolved in 100ml of distilled water, and stored in dark bottle at 4C°.

7- NBT solution (1.5 mM) 12.3mg of NBT dissolved in 100ml of distilled water and stored at 4C°.

## 4.4 Patients and blood samples:

Sixty samples of blood from leukemia and Hodgkin's disease (16 ALL, 15 CML, 10 CLL, 5 AML, 13 Hodgkin's), 7 with non malignant disease (control) and 6 normal individuals were collected from these patients, left for 30 minutes at room temperature, blood clots were separated at 3000 r.p.m for 10 minutes.

Sera were then aspirated and stored in sterilized tubes at -20 C° until time of analysis.

## 4.5 Determination of LDH activity in sera of leukemic and Hodgkin's disease.

1- Two ml of tris- NADH reagent is placed in 3ml cuvette.

2- Fifty µl of sera was added and the solution is mixed thoroughly, and then incubated for 10 minutes at 37 C°.

3- The reaction was initiated by the addition of 0.1 ml of working pyruvate solution. After mixing the cuvette is rapidly inserted into the spectrophotometer and the change in absorbance at 340 nm is immediately.

## Calculation:

The international units of activity (U) are expressed as micromoles of NADH per minute, and the enzyme concentrations are expressed as (U/L). The change in absorbance is related to international, units by the following equation for.

$$U/L = \Delta A/min \times \frac{2250\ \mu l}{50\ \mu l} \times \frac{1}{6220} \times 10^6$$

$$U/L = 7235 \times \Delta A/min$$

81

*Chapter Four* ————————————— *Determination of some biochemical constituents in sera of Leukemic and Hodgkin's patients*

Where:

2250 = the total volume μl.

50 = the serum volume used

6220 = molar extinction coefficient for NADH

$10^6$ = factor to convert concentration to μ mol/L.

Δ A = change in measured absorbance for time (t).

## 4.6 Determination of serum superoxide dismutase activity:

Superoxide dismutase (SOD) activity was estimated spectrophotrically by the method of Winterbourn et.al.,[287], with some modification, the method is based on the ability of the enzyme to inhibit the reduction of nitroblue tetra- zolium (NBT) by super oxide generated during the reaction of photoreduction riboflavin and oxygen.

## Procedure:

1- Addition of 0.2 ml of EDTA/ NaCN solution to 0.1 ml of serum, then 0.1ml of NBT solution were added.

2- The assay tubes were brought to a standard temperature (20-22C°), after that, 0.05 ml of riboflavin solution was added to each tube. The final assay volume of 3ml was made up with phosphate buffer 0.067 M, pH, 7.8.

3- Subsequent exposure to bright lighting was controlled by placing the assay tubes in a white- light box where they received uniform illumination for 20 minutes with 18w fluorescent tube attached to the lid, then the absorbance was read at 560 nm against distilled water.

4- To determine the control value, the absorbance for another set of tubes containing the same mixture was read at a 560 nm against distilled water, immediately after the addition riboflavin. (Riboflavin added after addition of buffer).

5- To determine SOD unit, ten tubes containing (10, 20, 40, 60, 80, 100, 200, 300, 400 and 500 μl) of normal serum samples, and another tube containing no serum were treated as described in the steps 1,2, and 3.

## Calculations:

1- Percentage inhibition was calculated from each absorbance in the presence and absence of the enzyme.

$$\text{Inhibition \%} = (A_E - A_{NE}) \times 100$$

where:

$A_E$ : The absorbance at 560 nm of the tubes containing different amounts of the enzyme.

$N_E$ : The absorbance at 560 nm in the absence of the enzyme.

2- The percentages of inhibition were plotted against the corresponding amount of serum (Figure 4-1).

3- SOD units were calculated from Figure (4-1) according to the amount of serum (V μl) which gives half the maximum inhibition of NBT redudtion (1 unit = 10.1 μl)

4- To calculate the SOD activity in sera of patients, the difference between absorbances before and after the light inhibition were, multiplied by the SOD unit.

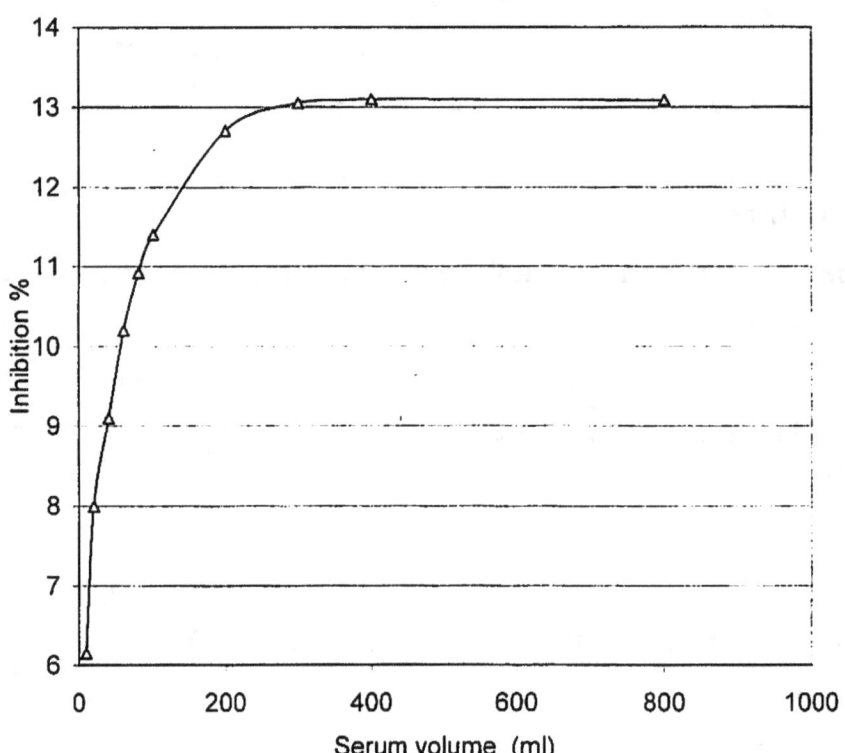

Figure (4-1)
The standard curve for determination of SOD unit.
All other details are explained in text.

## 4.7  Statistical methods:

Statistical analysis was performed by student's t-test[190].

## 4.8  IR spectra of sera of different types of Leukemic patients:

### 4.8.1  Instruments:

IR absorption spectra were recorded using KBr disc on-philips SP-3-300S Infrared spectrometric.

### 4.8.2  Blood sample preparation:

The Leukemic blood samples were classified in to three groups (ALL, CLL, CML) and one normal group. (Blood samples 5ml were obtained from individuals of all groups by vein puncture and the sera of all groups separated). The sera of all groups were lyophilized at the Al-Kindy Drugs Company.

**85**

*Chapter Four* ───────────── *Determination of some biochemical constituents in sera of Leukemic and Hodgkin's patients*

# Result and Discussion:

## Determination of LDH activity in sera of patients with leukemia, Hodgkin's and control:

The individual values of LDH activity for normal healthy controls and patients with leukemia, Hodgkin's are summarized in table (4-1).

Figure (4-2a) shows the distribution of the individual values of LDH activity in different types of leukemia, Hodgkin's patients. Figure (4-2b) shows the electrophoresis of LDH for leukemia, Hodgkin's patients.

LDH levels are elevated significantly in all group of leukemia, Hodgkin's patients compared to normal. All the five isozymes present in leukemia and Hodgkin's patients agree with the earlier reports[288-291].

It is observed in the present study that total LDH activity is found to be increased in-patients who had no response during, the course of chemotherapy. This could not be, useful in monitoring the response for treatment.

Patel, et al.,[194] found that LDH activity was significantly increased in all group of leukemic patients, and the patient, who successfully responded to the therapy show decrease in total LDH activity. Al-Mmudaffar, et al., reported an elevation in LDH activity in leukemic patients[292].

Elevation of LDH in malignancies is likely to be multifactorial , and no single explanation can be put forward to account for the whole spectrum of the malignancies. Increased serum LDH levels have been reported indifferent types of malignancies[167][293]-[295]. Other studied reported that LDH can be useful in monitoring the therapy of leukemic patients[236][237][240][244].

It is concluded that the total LDH activity could be as a biochemical marker for early assessment of response to therapy served[235].

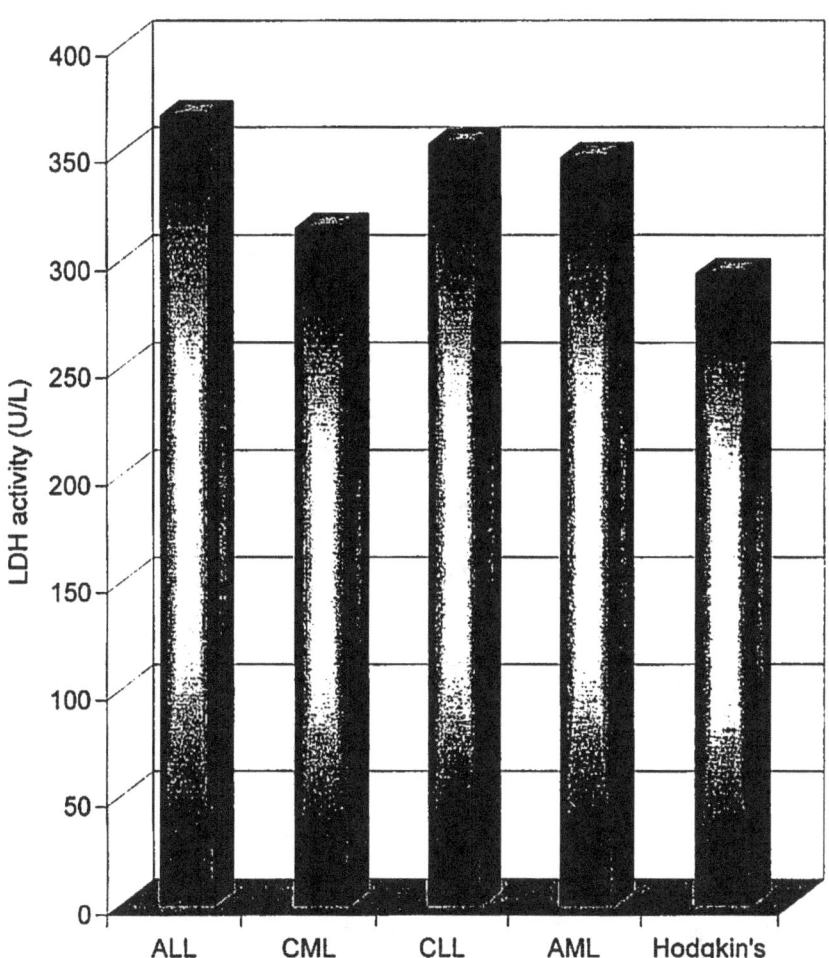

Figure (4-2a)
The mean of the individual values of LDH activity in sera of
leukemic  patients and Hodgkin's.
All other details are explained in text.

87

*Chapter Four* ———————————————— *Determination of some biochemical constituents
in sera of Leukemic and Hodgkin's patients*

Table (4-1)
Serum LDH activity in-patients with leukemia and Hodgkin's disease.
All details are explained in text.

| Group | N | Mean U/L | ± SD | T value | P value |
|---|---|---|---|---|---|
| ALL | 16 | 367 | 144.2 | 7.43 | 0.0000 |
| CML | 15 | 314.7 | 111.1 | 7.48 | 0.0000 |
| CLL | 10 | 353.7 | 127 | 6.32 | 0.0001 |
| AML | 5 | 347.4 | 128.3 | 4.31 | 0.0013 |
| Hodgkin's | 12 | 294 | 119.7 | 5.38 | 0.0002 |

The serum LDH activity for normal individuals = 126.3 ± SD 45.4U/L.

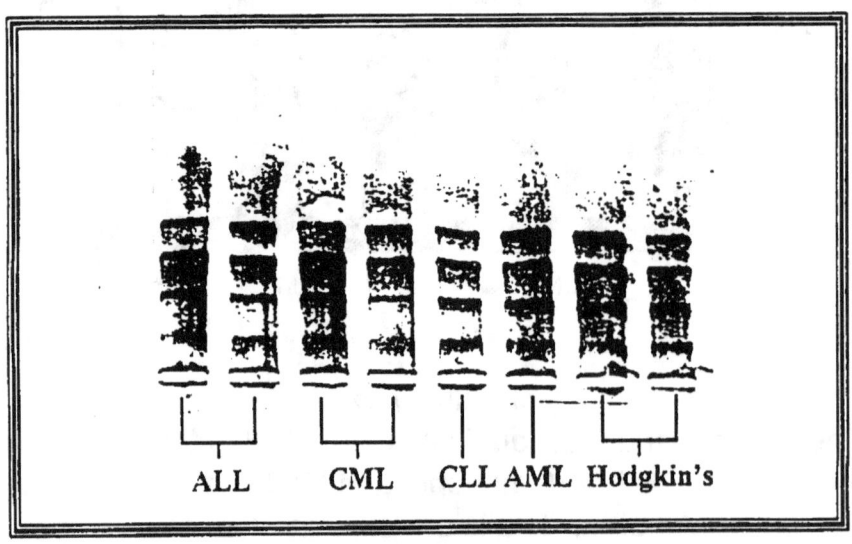

ALL    CML    CLL AML Hodgkin's

Figure (4-2b)
The electrophoresis of LDH for leukemic and Hodgkin's patients

## Determination of superoxide dismutase activity in sera of patients with leukemia, Hodgkin's and controls:

Superoxide dismutase activity was determined in sera of leukemic, Hodgkin's and controls patients. SOD activity was measured by the method of Winterbourn et al.,[287].

Table (4-2) shows the individual values of SOD activity in different types of leukemia, and Hodgkin's patients. The results revealed that the SOD activity decreased in leukemic, Hodgkin's patients compared to that of normal individuals and control.

No reliable differences were found in the SOD activity between the different types of leukemia, Figure (4-3) shows the distribution of the individual values of SOD activity in leukemia, Hodgkin's, and control patients, whereas a slight increase in SOD activity for Hodgkin's compared with that of leukemic, patients. Leukemic patients show higher level of LASA compared with normal healthy and controls patients. No statistical difference was found between different types of leukemia, and Hodgkin's patients.

It has been reported that there is a relationship between SOD activity and cancer, Knee, et al.,[296] and Bolzan et al.,[297] reported a lower levels of SOD in various malignant tumors compared with that of normal cells, also Fernandez has found that high metastatic cell lines contain less SOD than low metastatic cell ones[298].

Tumor cells have been shown to produce super oxide free radicals, if the rate of production of superoxide ion in tumor mitochondria is comparable to that found in the mitochondria from normal tissue, then the loss of SOD would result in net increase in the level of superoxide ion in the tumor cell[299].

Table (4-2)

Serum SOD activity in-patients with leukemia and Hodgkin's disease. All details are explained in text.

| Group | N | Mean | ± SD | T value | P value |
|-------|---|------|------|---------|---------|
| ALL | 12 | 1.4017 | .1073 | -8.02 | 0.0000 |
| CML | 8 | 1.3800 | .1166 | -6.55 | 0.0003 |
| CLL | 7 | 1.3086 | .0979 | -9.23 | 0.0001 |
| AML | 5 | 1.2480 | .0642 | -14.00 | 0.0002 |
| Hodgkin's | 8 | 1.4375 | .1316 | -4.57 | 0.0027 |
| Control | 7 | 1.828 | .2241 | 1.77 | 0.1300 |
| Normal | 6 | 2.01 | .230 | --------- | --------- |

Differences in the serum activity of SOD in-patients with cancer disease are probably due to the decreased enzyme level in the tumor cell. Another possibility is a decrease in the synthesis and release of SOD from the blood cells because of the malignant immune deficiency[296].

The values of serum SOD activity obtained from this assay were correlated with LASA concentrations in sera of leukemia, Hodgkin's and control patients. The results disclosed that, there is a significant negative between the two parameters. Consequently, the ratio of LASA/ SOD was calculated for all group of patients. Table (4.3) shows the value of LASA/ SOD ratio for all groups of patients and healthy individuals. The LASA/ SOD ratio for all types of leukemia is estimated to be about three fold higher, also in Hodgkin's was about two fold higher compared with normal healthy individuals.

The LASA/ SOD ratio was (27.69 ) in ALL and (30.6 ) in CML, (27.17 ) in CLL, (36.18 ) in AML leukemia, and (21.71 ) in Hodgkin's, the ratio was significantly higher in leukemia and Hodgkin's than in normal (P<0.001).

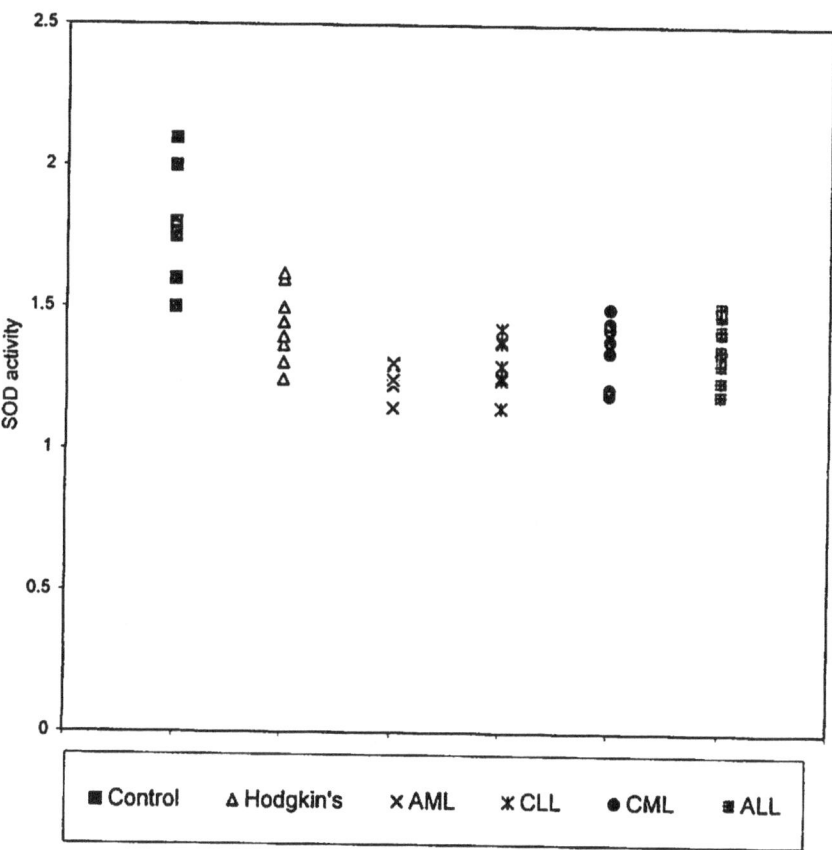

Figure (4-3)

Distribution of the inividual values of SOD activity in sera of leukemic patients, Hodgkin's, control and normal healthy.

Table (4-3)
Serum SOD activity, LASA and LASA/SOD ratio in-patients with
leukemia and Hodgkin's disease. All details are explained in text.

| Group | SOD activity | | LASA mg/dl | | LASA/SOD | |
|---|---|---|---|---|---|---|
| | Mean | ± SD | Mean | ± SD | Mean | ± SD |
| ALL | 1.401667 | 0.107266 | 38.58333 | 7.089792 | 27.69583 | 7.171079 |
| CML | 1.38 | 0.116619 | 40.5 | 12.27076 | 30.61625 | 11.42698 |
| CLL | 1.308571 | 0.097882 | 35 | 7.071068 | 27.17571 | 7.453021 |
| AML | 1.248 | 0.064187 | 44.6 | 13.16435 | 36.18 | 12.47546 |
| Hodgkin's | 1.4375 | 0.131557 | 30.375 | 10.59565 | 21.71875 | 9.695661 |
| Control | 1.82875 | 0.222418 | 28.625 | 6.926914 | 15.275 | 5.139969 |

Serum levels of SOD and LASA reflect the changes in the content of membrane glycolipids and cellular activity of SOD. Some authors admit that these changes in the tumor cell are in closed connection; the membrane of the neoplastic cell has an altered lipid content and structure organization, which leads to decreased antioxidant protection[300].

The loss of cell differentiation leads to an increase of the cell glycolipids, on other hand, a decrease in the intracellular SOD[301]. Figure (4.4) shows significant negative correlation between the two parameters in patients with leukemia, Hodgkin's, and it is also indicates that patients with high levels of LASA had low SOD content in the serum.

It has been reported that such negative correlation does not appears in children with non- cancer disease. This is in accordance with observation that changes in membrane glycolipids and cellular antioxidants occur in malignant tissues[302]. These findings suggest that SOD and LASA are good tumor markers.

92

*Chapter Four* ——————————— *Determination of some biochemical constituents*
*in sera of Leukemic and Hodgkin's patients*

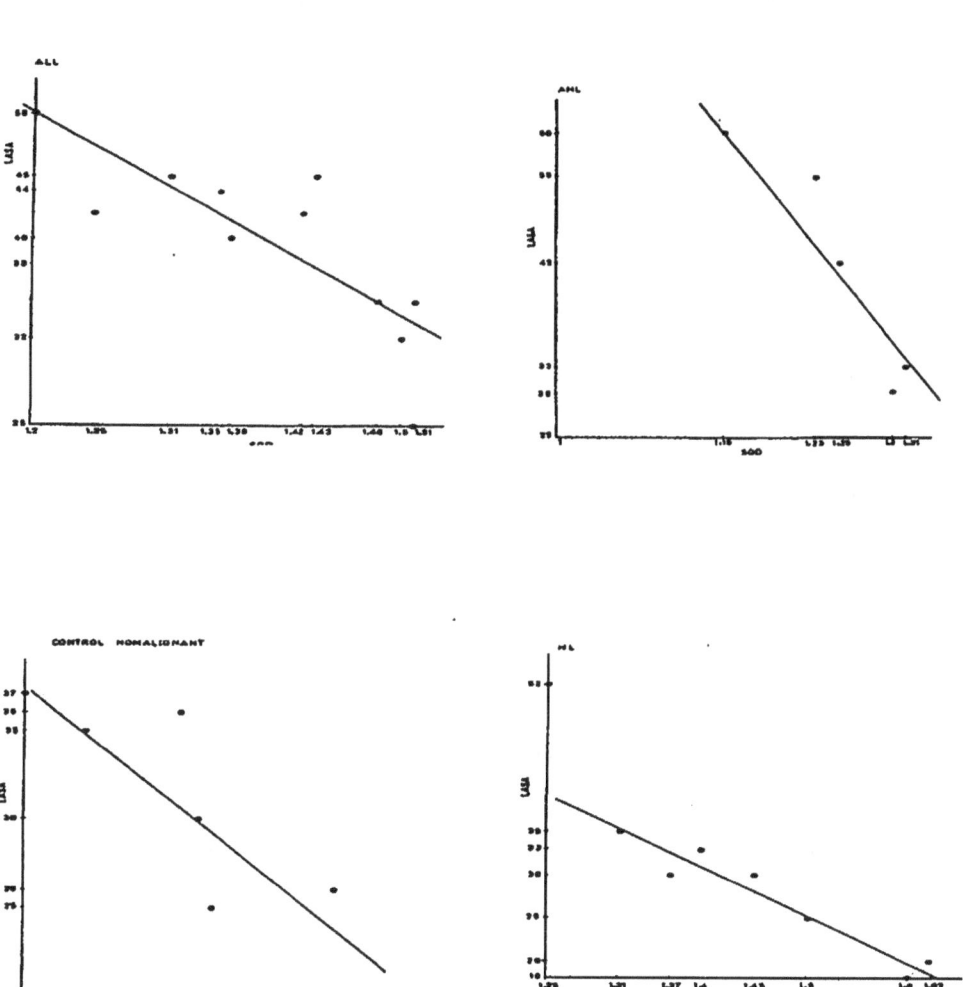

**Figure (4-4)**
**Correlation between SOD activity and the levels of LASA**
**in sera of leukemic, Hodgkin's and control patients.**
**All details are explained in the text.**

## Determination of sera trace elements and electrolytes:

The individual values of trace elements and electrolytes for the leukemia, Hodgkin's and pathological controls patients were determined by using the atomic absorption spectrophotometry. Figure (4.5) shows the individuals levels of some trace elements and electrolytes for Leukemic, Hodgkin's and control patients, the results show that there is an elevation in Cu, Fe level and Cu/ Zn ratio in sera of leukemic patients, also there is an increased in Cu level of Hodgkin's patients

On the other hand there is a reliable decrease of Zn, levels in sera of leukemic and Hodgkin's patients, compared to control individuals.

The decrease in Zn, levels and the increase in Cu level and Cu/ Zn ratio, indicate that there is a relation ship between these three parameters.

The Mg, Ca levels in sera of leukemia and Hodgkin's patients remain in normal levels, but 11% of leukemia had hypocalicamia and 15% of leukemia had hypercalicimia. The result shows that there is abnormal increase in ferrous level in some patients with CML before death.

Elliot et al., show that there is an increase in ferritin level of tumor tissue in 27% of breast cancer patients. Some authors, found that the abnormal increase in ferrous differences can indicate increase in cell numbers and its effect on growth of cancer cells in breast[303].

It is assumed that the decrease in zinc levels is a result of migration of zinc to the liver. Many authors indicated that the transfusion of zinc would help against spreading of cancer. On the other hand the effect of an increase in copper is not yet known[304].

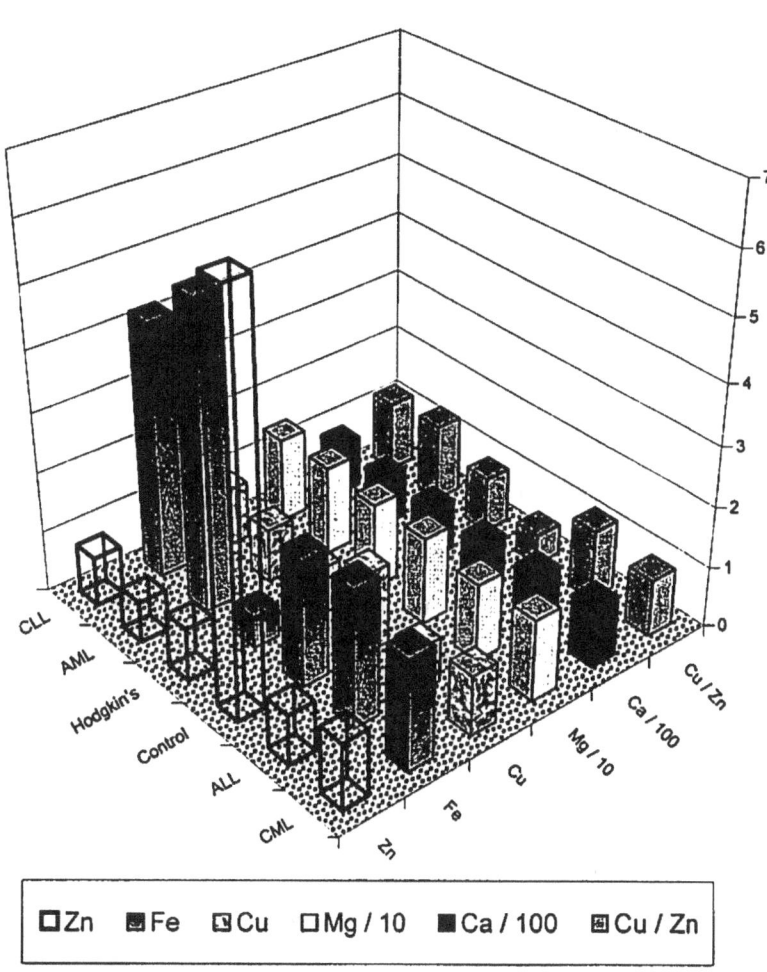

Figure (4-5)
The mean of the individual values of some trace elements
and some electrolytes in sera of leukemic patients,
Hodgkin's and control.
All other details are explained in text.

95

*Chapter Four* ————————————— *Determination of some biochemical constituents
in sera of Leukemic and Hodgkin's patients*

The increase of copper to zinc ratio in serum of ovarian cancer patients demonstrates the relationship between those two elements. This relationship can be utilized as indicator of the nature of tumor or it can be used for the diagnosis. It is thought that copper to zinc ratio in serum of cancer patients can be used as a indicator for the response of cancer treatment[305].

Cavallo et al., Y Ueel, et al., show that there is a relationship between cancer and copper to zinc ratio and show that, there is increase in this ratio compared with that of control group.

## IR Spectra of leukemic sera:

Infrared spectroscopy has been used in clinical laboratories. The infrared analysis of serum and other fluids from healthy individuals and from patients with various diseases including cancer[306]. IR spectra of leukemic groups were carried out in this work The Spectra showed many absorption bands of stretching bending and other vibration of different groups. The frequencies and the intensities of the absorption bands are shown in figure (4-6). As shown in figure (4-6), the spectra of the three groups shows absorption bands in the same region in the IR spectra of normal individual. Although several differences include:

1. a decrease in the frequency face about $10cm^{-1}$ in ALL, CLL & CML leukemic sera. due to secondary amine and primary a mines in the another side chains of proteins.

2. Increase in frequency of about 40-50 $cm^{-1}$ in the band of CLL, CML, ALL, from normal sera due to N-H stretching vibration of the - (NH) of the secondary amide in the polyamide backbone in the proteins and the ($NH_3$) of the amines salts in the side chains in the proteins.

96

*Chapter Four* ————————————— *Determination of some biochemical constituents in sera of Leukemic and Hodgkin's patients*

3. Strong bands at (3525 - 3250 cm$^{-1}$) is due to (0 - H) stretching vibration of and intermolecular hydrogen bonded of such as non-estrified cholesterol, the hydroxyl group, of Ser and Thr) residues in the protein molecules, and the hydroxyl groups of carbohydrate (glucose molecules). This band is also due to (H-H) stretching vibration of the primary amines in (Lys, Arg) residues, primary amide in (Asp, Clu) residues in the protein molecules.

4. Strong shoulder at 3350 cm$^{-1}$ in the ALL, 3400 cm$^{-1}$ in normal, 3500 cm$^{-1}$ in CML is due to the (N-H) stretching vibration of the hydrogen bonded secondary amide (NH) in the polyamide backbones in the protein molecules. This band is also due to the (H-H) stretching vibration of ammonium group (NH$_3$) of side chase and terminal amino groups in protein molecules.

5. Weak combination of overtone bands appears at (2000 - 1780 cm$^{-1}$) is due to the (C-C) stretching vibration of the aromatic ring in the side chain of proteins and other serum compounds.

6. Weak shoulder at 1720 cm$^{-1}$ is due to the C=0 stretching vibration of the ester groups of phosphotides (esterified cholesterol triacyl glyceride) and unioxid carboxyl groups in phospholipid acids.

7. Strong band at (1640 - 1620 cm$^{-5}$) is due to (C=0) stretching vibration of the amide band (amide I band) of the poly amide backbone of the proteins, (C=C).

97

*Chapter Four* ————————————— *Determination of some biochemical constituents
in sera of Leukemic and Hodgkin's patients*

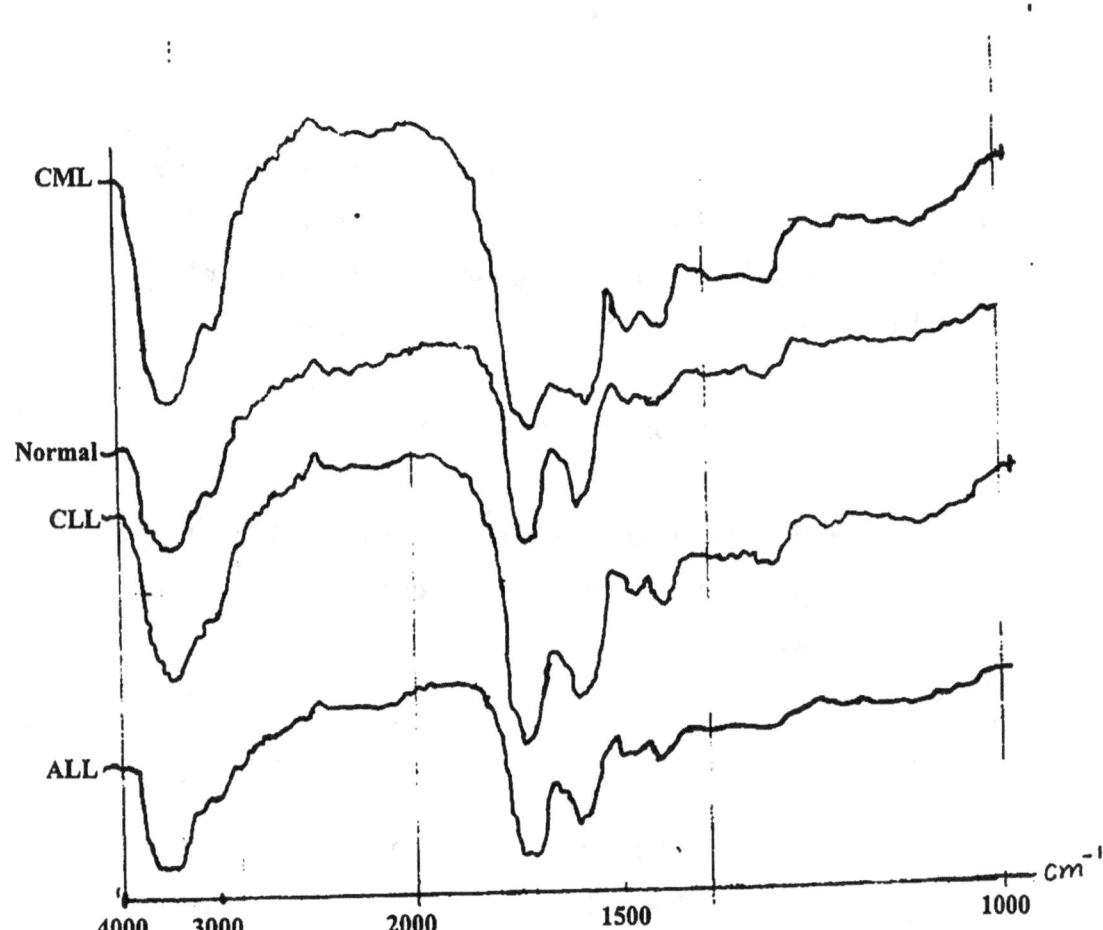

**Figure (4-6)**
**Infrared spectra of leukemic sera (ALL, CML & CLL)**
**and sera of normal individuals**

# CHAPTER FIVE

Binding characteristic studies of lectin of sera and bone marrows of Leukemic patients with Glycoprotein of human Red blood cells

**99**

*Chapter Five —— Binding characteristic studies of lectin of sera and bone marrows of Leukemic patients with Glycoprotein of human red blood cells*

Lectin possess a remarkable ability to agglutinate erythrocytes and other types of cells[307]. The ability of lectin to agglutinate red blood cells, makes easy detection. Lectins exhibit other interesting and unusual biological and chemical properties. Some of the lectins are specific in their reactions with human blood group (ABO) and subgroups (A$_1$) and therefore been used in blood typing and in investigations of chemical basis of blood group specificity[308].

Certain lectins are mitogenic, stimulate the transformation of lymphocytes from small "resting" cells into large blast- like cells which may ultimately undergo mitotic division. The stimulation of lymphocytes by lectins also provides an important tool in the examination of the biochemical events involved the conversion of a resting cell into actively growing one[309].

The interactions of lectins with cells can, in many instances, be inhibited specifically by simple sugars[310]. This finding has led to the conclusion that lectins bind specifically to saccharides on the surface of cells. Lectins also bind mono and oligosaccharides and specifically precipitate poly saccharides and glycoproteins, the precipitation is inhibited by sugars, as in the case of the agglutination reaction. However, few lectins with high specificity for sialic acid have been identified, such as limulin and carcinoscropine[311]. The *Escherichia coli* contains a lectin that binds D-mannose and its glycosides, and that presumably mediates bactretial attachment to cells. Another evidence for binding of *Vibrio cholera* to intestinal cell surfaces by a reaction inhibited by L- fucose has been presented[312]. Although a specific cellular function cannot yet be

**100**

*Chapter Five ——— Binding characteristic studies of lectin of sera and bone marrows of Leukemic patients with Glycoprotein of human red blood cells*

unequivocally assigned to any lectin. Evidences indicate that lectins have been adapted for avariety of cell surface and inter cellular functions in which the specific carbohydrate- binding site of the lectin binds a complementary saccharide- containing substance as a prelude to one of a number of biological actions[312].

Lectin which binds sialic acid residue of glycoproteins has also been isolated from wheat germ[313]. Sialic acid occupies an outstanding position both sterically and with respect to biological functions in various glycoproteins and glycolipid[314]. A lectin from human normal and cancerous tissues have been purified by conventional chromatography procedures[100][315]. Then characterized, and proves to be applicable to this study.

In this part an attempt was carried out to develop binding assay for lectin isolated from bone marrow of ALL, and blood of leukemia and Hodgkin's patients, and to characterize the binding of specific lectin with glycoprotein of human red cell by different techniques.

101

*Chapter Five ———— Binding characteristic studies of lectin of sera and bone marrows of Leukemic patients with Glycoprotein of human red blood cells*

# Materials and methods

## 5.1 Chemicals:

All chemicals and reagents mentioned in the section (2-1) of chapter two were used in the experiments of this chapter.

## 5.2 Instruments:

All instruments that described in section (2-2) of chapter two were used in the following experiments of this chapter.

## 5.3 Buffers and Reagents:

All buffer solutions are prepared by dissolving the appropriate amount of salts in distilled water and the required pH was adjusted. The following stock solutions are prepared in the experiments of this chapter.

- 0.02 M HCl.

- 0.02 M Tris(hydroxymethylamino methane,2.973gm in 1000ml distilled water).

- 0.075 M $Na_2HPO_4$(8.98 gm in 1000 ml distilled water).

- 0.075 M $KH_2PO_4$(13.25 gm in 1000 ml distilled water).

- 0.02 M $CaCl_2$ (2.12 gm in 1000 ml of buffer).

- 0.15 M NaCl (8.765 gm in 1000 ml of buffer).

- 0.002 M EDTA (0.75 gm in 1000ml of buffer).

- 0.15 $MnCl_2$ (2 gm in 1000ml of buffer).

- 0.01 $ZnCl_2$ (1.36 gm in 1000 ml of buffer).

- 0.01 M NaF (0.4 gm in 1000 ml of buffer).

- 0.02 M $MgCl_2$(1.9 gm in 1000 ml of buffer).

- 0.02 M KI (3.2 gm in 1000 ml of buffer).

102

*Chapter Five ——— Binding characteristic studies of lectin of sera and bone marrows
of Leukemic patients with Glycoprotein of human red blood cells*

**Working buffer:**

Assay buffer (Tris- Saline buffer pH 8). Each liter of 0.02m tris- HCL pH containing 0.15 M NaCl and 0.02 CaCl$_2$.

## 5.4 Bone marrow collection and preparation:

Human leukemic bone marrow was removed from patients with All leukemia, diagnosed by specialists. The bone marrow washed with Tris buffer and centerifugated at 4000 r.p.m for 15 minutes. The supernatant was used as a source of lectin for the binding expressed as hemagglutination. The supernatant stored at- 20 $\overset{\circ}{C}$ till time of assay.

## 5.5 Determination of protein concentration (Lectin) from bone marrow and blood of leukemia and Hodgkin's patients

Protein was determined by the method of Lowry. et al., using bovine serum albumin as a standard[188].

## 5.6 Preliminary test for the lectin binding to erythrocyte surface glycoconjugates:

### 5.6.1 Preparation of standard erythrocytes suspension for hemagglutination:

Fresh human standard erythrocytes (type B) were washed (3-4) times with an appropriate amount of normal saline (0.9% NaCl solution), then the cells were diluted with normal saline to give an absorbance value about 2 at 620nm[316]. The suspension was prepared on the day of the assay.

103

*Chapter Five* ——— *Binding characteristic studies of lectin of sera and bone marrows of Leukemic patients with Glycoprotein of human red blood cells*

## 5.6.2 Procedure of lectin binding to glycoprotein:

The total binding of lectin from (bone marrow and blood of leukemia and Hodgkin's patients) to erythrocyte surface glycoconjates was preliminary tested by the hemagglutination assay:

This assay is a modification of the method described by Liener[317][318]. The assay includes the following steps:

1- Half ml of erythrocyte suspension was added to $50\,\mu l$ of lectin from sera and bone marrow (leukemic lectin), then $450\,\mu l$ of Tris- HCL buffer (pH,8) were added to give a final volume of 1ml. The assay mixture was incubated for 30 minutes at room temperature.

2- The assay mixture was centrifuged for 5 minutes at 2000 r.p.m, then the supernatant was removed.

3- The remaining red blood cells were resuspended in $500\,\mu l$ of Tris HCl buffer (pH,8). The reacted (bound) cells were allowed to sediment for 30 minutes, then the absorbance at 620 nm of the upper layer (free lectin and cells) of the assay solution was measured.

## Calculation:

Total binding represents the amount of lectin which binds the erythrocytes surface glycoconjugate causing hemagglutination.

$$TB\% = \frac{A - A^*}{A} \times 100$$

Where: TB%: The percent of total binding of lectin to erythrocyte surface glycoconjugates.

A: The absorbance of standard erythorocyte suspension at 620 nm.

A*: The absorbance of free (unbound) erythrocyte at 620 nm.

104

*Chapter Five ——— Binding characteristic studies of lectin of sera and bone marrows*
*of Leukemic patients with Glycoprotein of human red blood cells*

## 5.6.3 Determination of non- specific binding of leukemic lectin from (sera and bone marrow) to erythrocyte surface glycoconjugates:

The erythrocyte suspension used in this assay was prepared as follows:

1- The erythrocytes were washed (3-4) times with normal saline (0.9% NaCl), then two times with assay (Tris- HCl buffer pH, 8) and the suspension was prepared.

2- Hundred μl of neuraminidase (500 unit/ml) were added to 10ml of the suspension. The mixture was shaken for four hours at 25 $\overset{\circ}{C}$.

3- After centrifugation, the supernatant was removed and the remaining red blood cells were diluted with an appropriate amount of normal saline to give an absorbance value of about 2 at 620 nm and then the same steps described section (5.6.1.2) were followed to determine the percent of the non- specific binding.

## Calculations:

The percent of non- specific binding was calculated using the following equation:

$$NSB\% = \frac{A^{\bullet} - A^{*}}{A^{\bullet}} \times 100$$

Where:

NSB%: The percent of non- specific binding.

$A^{\bullet}$: The absorbance of neuraminidase- treated erythrocyte suspension at 620nm.

$A^{*}$: The absorbance of free (unbound) erythrocytes at 620 nm.

105

*Chapter Five*————*Binding characteristic studies of lectin of sera and bone marrows*
*of Leukemic patients with Glycoprotein of human red blood cells*

### 5.6.4 Determination of the specific binding of lectin (from sera and bone marrow) to erythrocyte surface glycoconjugates:

The percent of specific binding of lectin (from sera and bone marrow) to glycoconjugates was calculated by subtracting of non- specific binding from the percent of the total binding:

$$SB\% = TB\% - NSB\%$$

Where:

### 5.6.5 Factors effecting lectin binding to erythrocyte surface glycoconjugates in leukemic and Hodgkin's disease

### 5.6.5.1 The effect of different amount of lectin (sera and bone marrow) on its binding to erythrocyte surface glyconjugates

Half ml of erythrocyte suspension, was added, to 100 µl of increasing amount (200, 600, 1000, 1400, 1800, 2200, 2600, 3000, 3400 and 4000 µg) of lectin from blood and (50, 75, 100, 125, 150, 175, 200, 225, 250 and 400 µg) from bone marrow of leukemic and Hodgkin's patients, then the total volume of 1ml (completed with Tris-HCl buffer pH,8). After incubation for 30 minutes at 25 $\overset{\circ}{C}$, the step 2 and 3 described in section (5.6.2) were followed.

### Calculation:

1- The same equation mentioned in section (5.6.2) was used to calculate the percent of total binding.

2- The percent of specific binding (SB%) was calculated by using the equation mentioned in section (5.6.3).

3- The percent of specific binding (SB%) was plotted against the corresponding protein amount include in each mixture as shown in figure (5-1).

106

*Chapter Five——— Binding characteristic studies of lectin of sera and bone marrows*
*of Leukemic patients with Glycoprotein of human red blood cells*

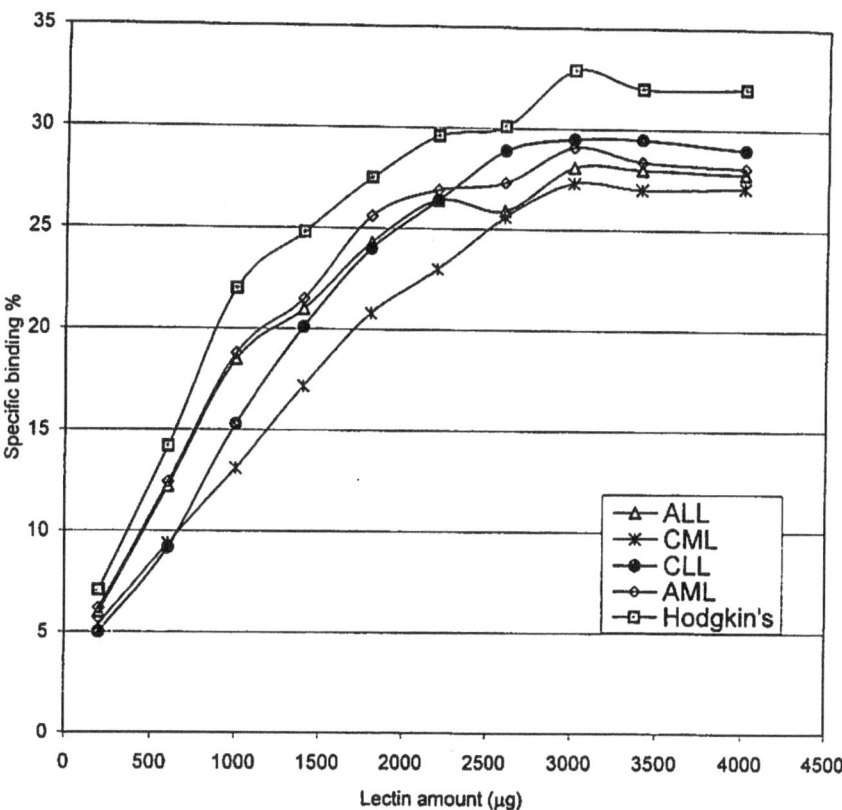

Figure (5-1)
Effect of cancerous lectin's amount on the binding of
lectin to erythrocyte surface glyconconjugates.
All other details are explained in text.

107

*Chapter Five* ——— *Binding characteristic studies of lectin of sera and bone marrows of Leukemic patients with Glycoprotein of human red blood cells*

## 5.6.5.2 The effect of pH on lectin (sera and bone marrow) binding to erythrocyte surface glycoconjugtes:

Half ml of erythrocyte suspension was added to 50 µl of lectin (3000 µg) and (200 µg) from blood and bone marrow of leukemia and Hodgkin's patients, the volume of mixture were made up to 1ml, with Tris HCl buffer of different pH values (7, 7.5, 8, 8.5, 9) the assay tubes were incubated for 30 minutes at 25 $\overset{\circ}{C}$, after that the step 2 and 3 of section (5.6.2) were repeated.

**Calculation:**

1- The same equation mentioned in experiment (5.6.2) was used to calculate the percent of total binding.

2- The percent of specific binding (SB%) was calculated by using the equation mentioned in section (5.6.3).

3- The percentage of specific binding (SB%) was plotted against the corresponding pH values, as shown in Figure (5-2).

## 5.6.5.3 The effect of temperature on the binding of lectin (sera and bone marrow) to erythrocyte surface glycoconjugates

Half ml of erythrocyte suspension was added to 50 µl of lectin (3000 µg) and (200 µg). The final assay volumes were made up to 1ml with Tris-HCl buffer (pH,8.5). The tubes were incubated for 30 minutes at different temperatures (5, 15, 25, 30, 37 and 40 C°). After that the steps 2 and 3 of section (5.6.2) were repeated.

**Calculations:**

1- The same equation mentioned in experiment (5.6.2) was used to calculate the percent of total binding.

2- The percent of specific binding (SB%) was calculated by using the equation mentioned in section (5.6.3).

3- The percentages of specific binding (SB%) were plotted against their corresponding temperature as shown in figure (5-3).

108

*Chapter Five* ———— *Binding characteristic studies of lectin of sera and bone marrows of Leukemic patients with Glycoprotein of human red blood cells*

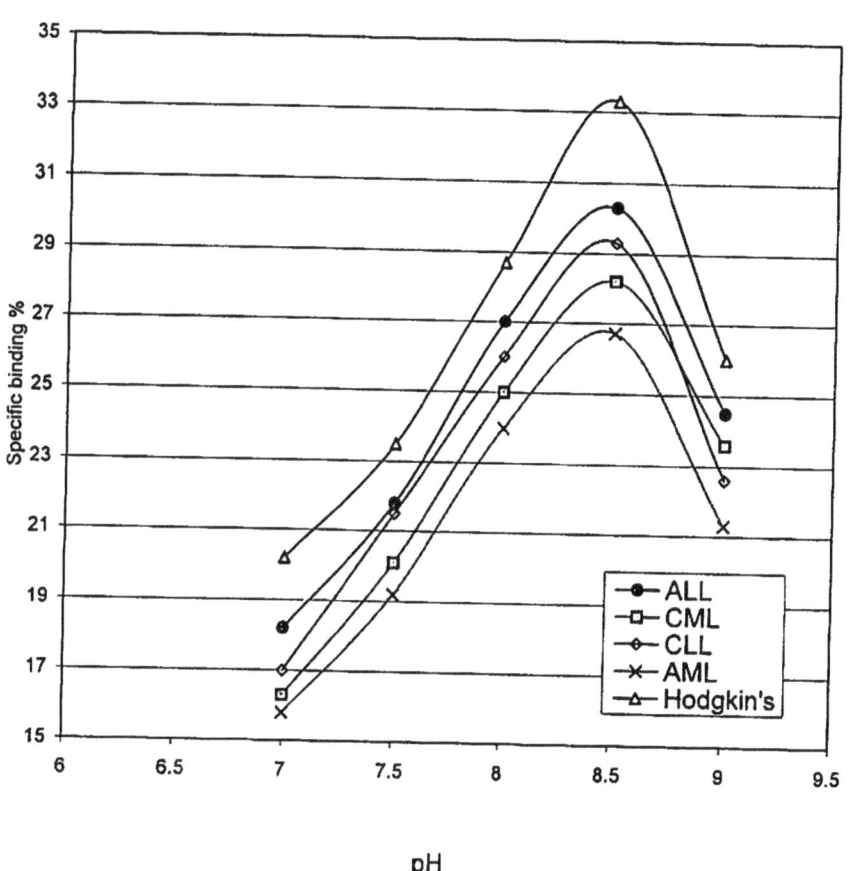

Figure (5-2)
Effect of pH on the binding of lectin to erythrocyte
surface glyconconjugates.
All other details are explained in text

109

*Chapter Five* ——— *Binding characteristic studies of lectin of sera and bone marrows
of Leukemic patients with Glycoprotein of human red blood cells*

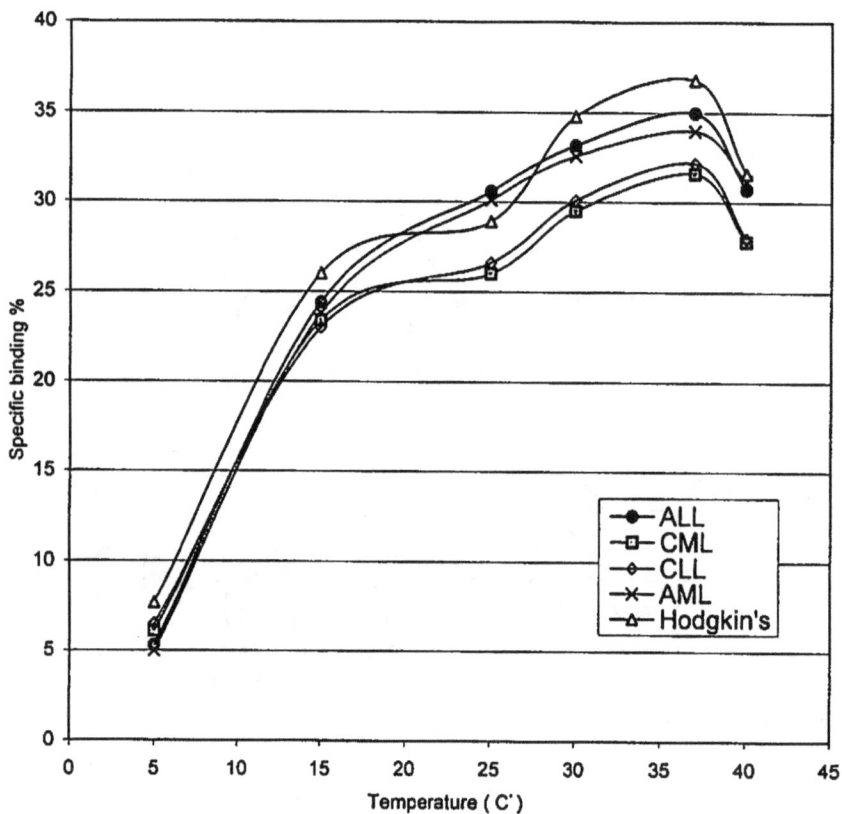

Figure (5-3)
Effect of Temperature on the binding of lectin to
erythrocyte surface glyconconjugates.
All other details are explained in text.

110

*Chapter Five ——— Binding characteristic studies of lectin of sera and bone marrows of Leukemic patients with Glycoprotein of human red blood cells*

## 5.6.5.4 The effect of incubation time on lectin (sera and bone marrow) binding to erythrocyte surface glycoconjugates

Half ml of erythrocyte suspension was added to $50\,\mu l$ of lectin $(3000\,\mu g)$, and $(200\,\mu g)$, the final volume was made up to 1ml with Tris-HCl buffer (pH,8.5). The assay tubes were incubated at $37\,C°$ for (15, 30, 60, 90 and 120 minutes) then the step 2 and 3 of section (5.6.2) were repeated.

**Calculation:**

1- The same equation mentioned in experiment (5.6.2) was used to calculate the percent of total binding.

2- The percent of specific binding (SB%) was calculated by using the equation mentioned in section (5.6.3).

3- The percentages of specific binding (SB%) were plotted against their corresponding times, as shown in figure (5-4).

## 5.6.5.5 The effect of exogenous $Ca^{+2}$ concentration on the binding of lectin (sera and bone marrow) to erythrocyte surface glycoconjugates:

Half ml of erythrocyte suspension was added to $50\,\mu l$ of lectin $(3000\,\mu g)$ and $(200\,\mu g)$, the final assay volume was made up to 1ml with Tris-HCl buffer (pH, 8.5) which contain different $Ca^{+2}$ concentration (5,10,15,20,30mM). The assay mixtures were incubated for 90 minutes at $37\,C°$. After that the steps 2 and 3 of section (5.6.2) were repeated.

**Calculation:**

1- The same equation mentioned in experiment (5.6.2) was used to calculate the percent of total binding.

2- The percent of specific binding (SB%), was calculated by using the equation mentioned in section (5.6.3).

3-The percentages of specific binding (SB%) were plotted against their corresponding $Ca^{+2}$ concentrations, as shown in figure (5-5).

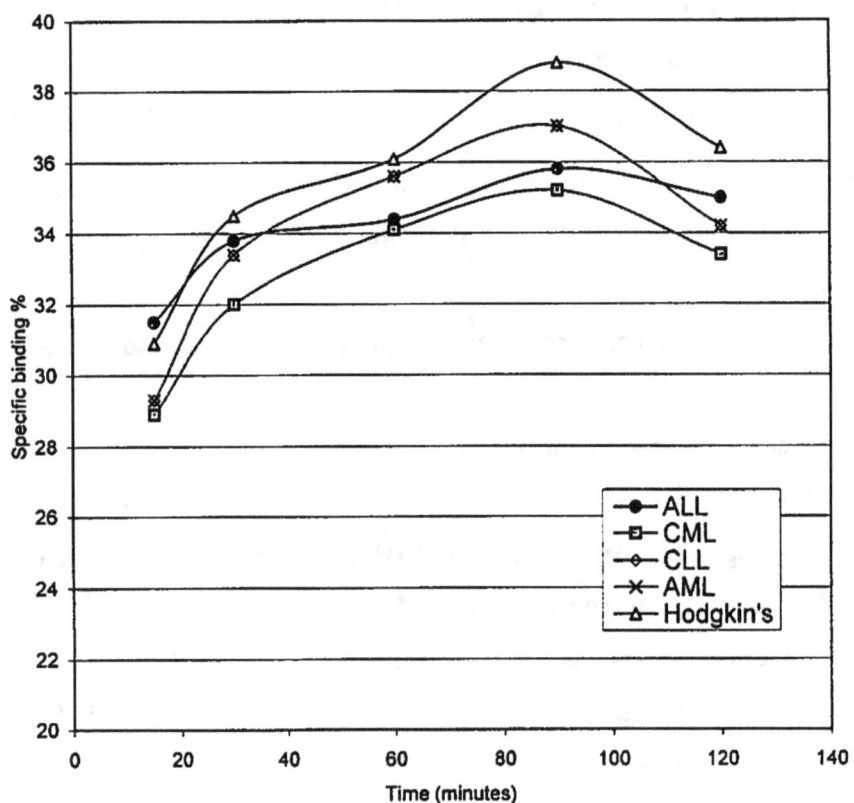

Figure (5-4)
Effect of incubation time on the binding of lectin to
erythrocyte surface glyconconjugates.
All other details are explained in text.

112

*Chapter Five ——— Binding characteristic studies of lectin of sera and bone marrows of Leukemic patients with Glycoprotein of human red blood cells*

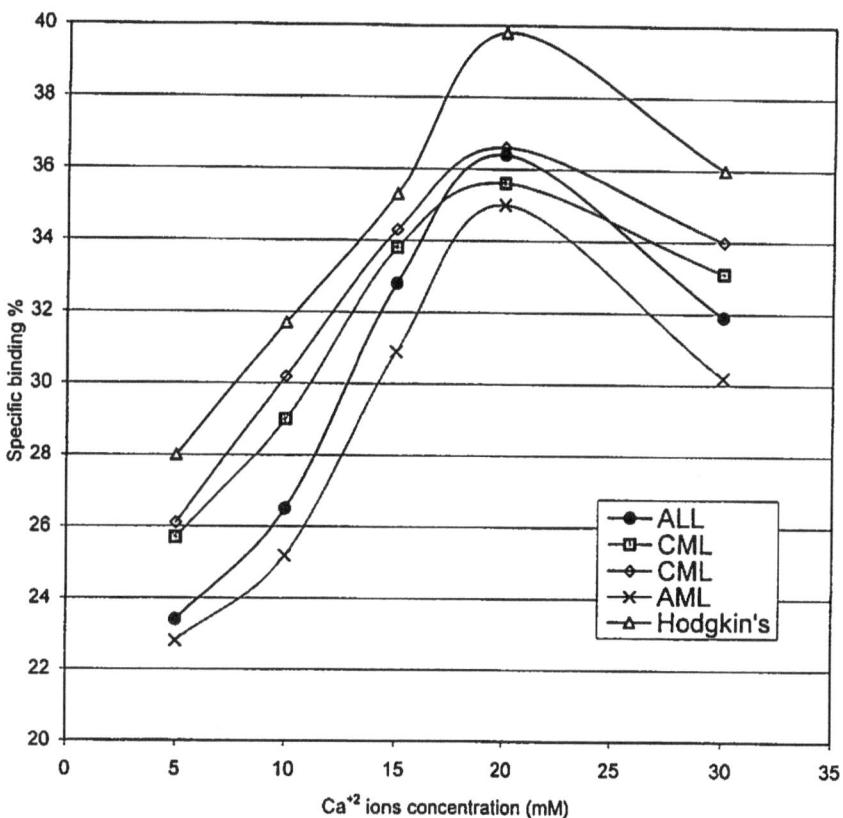

Figure (5-5)
Effect of Ca$^{+2}$ ions concentration on the binding of
lectin to erythrocyte surface glyconconjugates.
All other details are explained in text.

113

*Chapter Five* ——— *Binding characteristic studies of lectin of sera and bone marrows of Leukemic patients with Glycoprotein of human red blood cells*

**5.6.5.6 The effect of monovalent salts on lectin (sera and bone marrow) binding to erythrocyte surface glycoconjugate:**

Half ml of erythrocyte suspension was added to 50 μl of lectin (3000 μg), and (200 μg), the final volumes were completed to 1ml with Tris-HCl buffer (pH, 8.5) which contains different NaCl concentration (50,100,150,200 and 300 mM): After incubation for 90 minutes at 37 C° after that the steps 2 and 3 of section (5.6.2) were repeated. The same procedure was repeated for NaF, KI salts.

**Calculation:**

1- The same equation mentioned in experiment (5.6.2) was used to calculate the percent of total binding.

2- The percent of specific binding (SB%) was calculated by using the equation mentioned in section (5.6.3).

3- The percentages of specific binding (SB%) were plotted against their corresponding NaCl concentration, as shown in Figure (5-6).

**5.6.5.7 The effect of divalent salts on lectin (sera and bone marrow) binding to erythrocyte surface glycoconjugates:**

Half ml of erythrocyte suspension was added to 50 μl of lectin (3000 μg) and (200 μg), the volume of assay mixtures was completed with the Tris- HCl buffer (pH, 8.5) which contains the $MgCl_2$ concentrations (5, 10, 15 and 20mM). The tubes were incubated for 90 minutes at 37 C°, then the step 2 and 3 of section (5.6.2) were repeated. The same procedure was repeated for $MnCl_2$ and $ZnCl_2$ salts.

**Calculation:**

1- The same equation mentioned in experiment (5.6.2) was used to calculate the percent of total binding.

2- The percent of specific binding (SB%) was calculated by the same equation mentioned in section (5.6.3).

3- The percentages of specific binding (SB%) were plotted against their corresponding $Mg^{+2}$ concentrations as shown in Figure (5-6).

114

*Chapter Five* ———— *Binding characteristic studies of lectin of sera and bone marrows of Leukemic patients with Glycoprotein of human red blood cells*

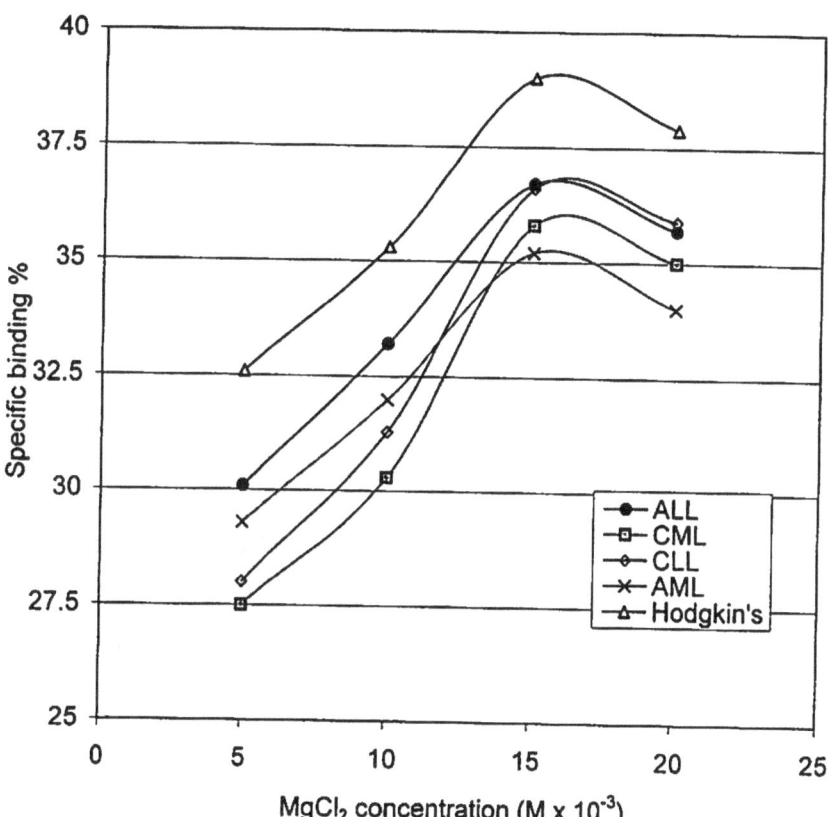

Figure (5-6)
Effect of MgCl$_2$ concentration on the binding of lectin
to erythrocyte surface glyconconjugates.
All other details are explained in text.

115

*Chapter Five* ———— *Binding characteristic studies of lectin of sera and bone marrows of Leukemic patients with Glycoprotein of human red blood cells*

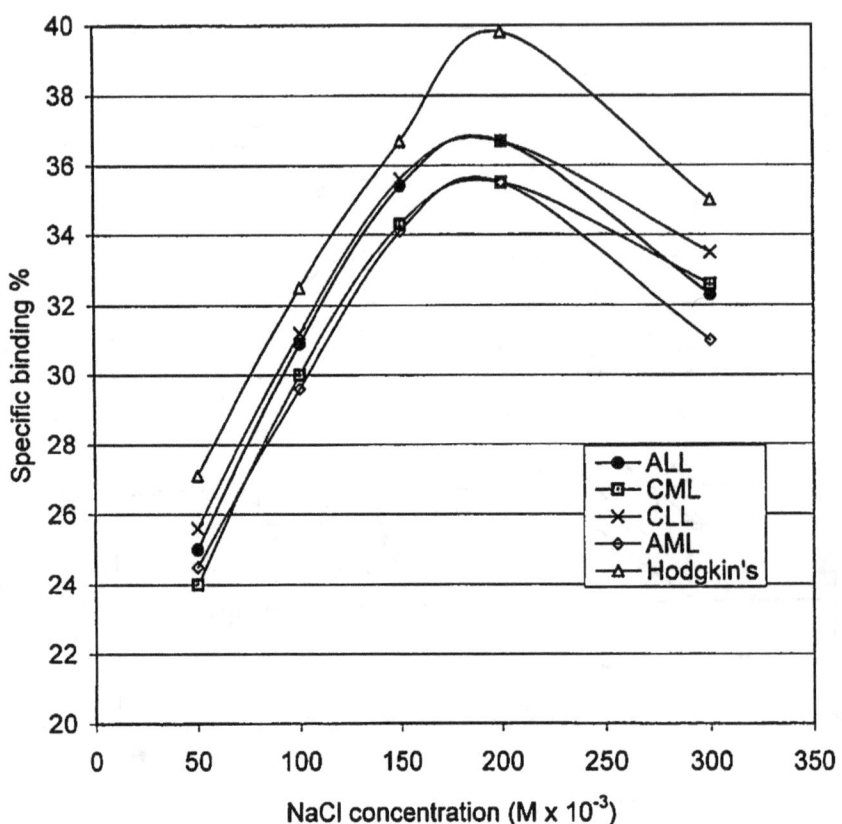

Figure (5-7)
Effect of NaCl concentration on the binding of
lectin to erythrocyte surface glyconconjugates.
All other details are explained in text

116

*Chapter Five* ——— *Binding characteristic studies of lectin of sera and bone marrows of Leukemic patients with Glycoprotein of human red blood cells*

## 5.6.5.8 The effect of different denaturating agents on lectin (sera and bone marrow) binding to erythrocyte surface glycoconjugates:

Half ml of erythrocyte suspension was added to $50\,\mu l$ of lectin $(3000\,\mu g)$ and $(200\,\mu g)$, the final volumes of assay mixture were completed with Tris- HCl buffer pH, 8.5 which contains different concentration of the following denaturing agents urea (3, 4, 5 and 6 M) polyethylene glycol (0.5%, 1% and 4%), sodium hydroxide solution (0.15M) and (0.15M) of hydrochloric acid solution. The tubes incubated for 90 minutes at $37C^{\circ}$. After that the steps 2 and 3 of section (5.6.2) was repeated.

## Calculation:

1- The same equation mentioned in experiment (5.6.2) was used to calculate the percent of total binding.

2- The percent of specific binding (SB%) was calculated by using the equation mentioned in section (5.6.3).

3- The percentage of specific binding (SB%) were summarized in table (5-1).

### Table (5-1): Effect of denaturating agents on the lectin (sera bone marrow) binding to erythrocyte surface glycoconjugates.
### All details are explained in text

| Groups | Types of test | Reagent added (M) | Specific binding % |
|--------|--------------|-------------------|--------------------|
| ALL | Urea | 3 | 79.1 |
|  |  | 4 | 71.4 |
|  |  | 5 | 63.5 |
|  |  | 6 | 47.2 |
| CML | Urea | 3 | 75.6 |
|  |  | 4 | 69.3 |
|  |  | 5 | 64.0 |
|  |  | 6 | 49.4 |

*Chapter Five ——— Binding characteristic studies of lectin of sera and bone marrows of Leukemic patients with Glycoprotein of human red blood cells*

| Groups | Types of test | Reagent added (M) | Specific binding % |
|---|---|---|---|
| Hodgkin's | Urea | 3 | 70.8 |
| | | 4 | 61.7 |
| | | 5 | 53.0 |
| | | 6 | 40.9 |
| ALL | PEG | 0.5% | 76.1 |
| | | 1 | 65.3 |
| | | 2 | 47.2 |
| | | 4 | 25.6 |
| CML | PEG | 0.5% | 74.6 |
| | | 1 | 66.3 |
| | | 2 | 48.2 |
| | | 4 | 25.0 |
| Hodgkin's | PEG | 0.5% | 63.2 |
| | | 1 | 41.8 |
| | | 2 | 30.5 |
| | | 4 | 22.9 |
| ALL | NaOH | 0.15 | 31.8 |
| | HCl | 0.15 | 20.4 |
| CML | NaOH | 0.15 | 28.3 |
| | HCl | 0.15 | 18.2 |
| Hodgkin's | NaOH | 0.15 | 27.6 |
| | HCl | 0.15 | 21.0 |

118

*Chapter Five* —————— *Binding characteristic studies of lectin of sera and bone marrows of Leukemic patients with Glycoprotein of human red blood cells*

## 5.6.5.9 Inhibition studies on lectin (sera and bone marrow) binding to erythrocyte surface glycoconjugates:

The inhibition studies on the binding of lectin to erythrocyte surface were performed by different inhibitors, such, as (sialic acid, glucuronic acid, fructose, mannose and Xylose). All experiments were carried out at optimum conditions, lectin amount (3000 µg) from blood, (200 µg) from bone marrow, temperature, 37C°, pH, 8.5, incubation time 90 minutes and $Ca^{+2}$ concentration, 20mM. Half ml of erythrocyte suspension was added to 50 µl of lectin. The assay volume was made up to 1ml with Tris- HCl buffer, pH, 8.5. After the addition of different concentration of (sialic acid, glucoronic acid, fructose, mannouse and xylose), then the assay tubes were incubated for 90 minutes at 37C°. After that, the step 2 and 3 of section (5.6.2) were repeated.

## Calculation:

1- The same equation mentioned in experiment (5.6.2) was used to calculate the percent of total binding before and after addition of inhibitor.

2- The percent of specific binding (SB%) was calculated before and after addition of inhibitor by using the equation mentioned in section (5.6.3).

3- The percent of inhibition was calculated as the difference between the percent of specific binding (SB%) in absence of inhibitor and that obtained in the presence of it. The data of inhibition was shown in table (5-2).

119

*Chapter Five ———— Binding characteristic studies of lectin of sera and bone marrows of Leukemic patients with Glycoprotein of human red blood cells*

**Table (5-2): Inhibition of binding of lectin to erythrocyte surface glycoconjugates. All details are explained in the text**

| Groups | Type of carbohydrate | Carbohydrate concentration (mM) | Inhibition % |
|---|---|---|---|
| ALL | D-glucuronic acid | 1 | 3.5 |
| | | 10 | 14.6 |
| | | 15 | 15.7 |
| | | 20 | 18.6 |
| CML | D-glucuronic acid | 1 | 2.5 |
| | | 10 | 12.9 |
| | | 15 | 14.8 |
| | | 20 | 17.1 |
| CLL | D-glucuronic acid | 1 | 3.1 |
| | | 10 | 13.4 |
| | | 15 | 15.2 |
| | | 20 | 17.0 |
| AML | D-glucuronic acid | 1 | 4.1 |
| | | 10 | 15.2 |
| | | 15 | 16.0 |
| | | 20 | 19.1 |
| Hodgkin's | D-glucuronic acid | 1 | 5.2 |
| | | 10 | 15.4 |
| | | 15 | 17.0 |
| | | 20 | 20.3 |
| ALL | Fructose | 30 | 5.9 |
| | Mannose | 30 | 5.5 |
| | Xylose | 30 | 4.0 |
| CML | Fructose | 80 | 5.5` |
| | Mannose | 30 | 5.2 |
| | Xylose | 30 | 4.5 |
| CLL | Fructose | 30 | 5.3 |
| | Mannose | 30 | 5.0 |
| | Xylose | 30 | 4.2 |

| Groups | Type of carbohydrate test | Carbohydrate concentration (mM) | Inhibition % |
|---|---|---|---|
| AML | Fructose | 30 | 6.0 |
| | Mannose | 30 | 5.4 |
| | Xylose | 30 | 5.0 |
| Hodgkin's | Fructose | 30 | 6.4 |
| | Mannose | 30 | 6.1 |
| | Xylose | 30 | 6.3 |
| ALL | Sialic acid | 1.0 | 7.6 |
| | | 1.5 | 10.5 |
| | | 2.0 | 15.2 |
| | | 2.5 | 18.1 |
| | | 3.0 | 23.4 |
| | | 3.5 | 28.3 |
| CML | Sialic acid | 1.0 | 7.0 |
| | | 1.5 | 9.7 |
| | | 2.0 | 14.5 |
| | | 2.5 | 17.6 |
| | | 3.0 | 22.2 |
| | | 3.5 | 25.6 |
| CLL | Sialic acid | 1.0 | 7.2 |
| | | 1.5 | 10.1 |
| | | 2.0 | 14.6 |
| | | 2.5 | 17.9 |
| | | 3.0 | 23.0 |
| | | 3.5 | 28.0 |
| AML | sialic acid | 1.0 | 7.4 |
| | | 1.5 | 10.2 |
| | | 2.0 | 15.3 |
| | | 2.5 | 18.2 |
| | | 3.0 | 23.3 |
| | | 3.5 | 26.6 |
| Hodgkin's | Sialic acid | 1.0 | 6.9 |
| | | 1.5 | 11.3 |
| | | 2.0 | 16.3 |
| | | 2.5 | 20.7 |
| | | 3.0 | 25.6 |
| | | 3.5 | 29.3 |

## 5.7 Purification of lectin from bone marrow:

### 5.7.1 Gel preparation and column packing:

a- Preparation of the column: the dimension of the column were chosen according the following equation[319]:

$$diameter = \sqrt[3]{m/10} \, (cm) \qquad\qquad ...(1)$$

where:

m = amount of protein in mg.

**Length = 30 x diameter**           ...(2)

b- Preparation of the gel: The gel (sephadex G 150) was allowed to swell in excess of 0.02 M Tris- saline buffer pH 7.2 containing 0.01M $CaCl_2$ and left to stand for three days at room temperature with out string, then the slurry was poured carefully into a vertical glass column down the wall using a glass rod. After the gel has settled, the column was equilibrated with Tris saline buffer pH, 7.2 for 24 hours with the dimension of 1.5 x 75 cm.

### 5.7.2 Void volume (Vo) Determination:

The volume of the gel was determined by using blue dextrin 2000 at (concentration of 1mg/ml in Tris- saline buffer pH, 7.2). One ml of blue dextrin solution was applied to the column surface carefully, the elution was carried out with the same buffer, using a flow rate of 8ml/ hour, fraction of three ml were then collected and their absorbance was measured at 600nm to determine the void volume.

122

*Chapter Five —— Binding characteristic studies of lectin of sera and bone marrows of Leukemic patients with Glycoprotein of human red blood cells*

### 5.7.3 Kav value Determination:

The Kav value was calculated for lectin, eluted, by using the following formula:

$$Kav = \frac{V_e - V_o}{V_t - V_o} \qquad ...(3)$$

where:

$V_o$ = Void volume.

$V_e$ = Elution volume of each protein (Lectin, Dextrin)

$V_t$ = Total gel- bed volume: determined from the following equation.

$$V_t = \left[\frac{\text{column diameter}}{2}\right]^2 \times \frac{22}{7} \times \text{column length} \qquad ...(4)$$

### 5.7.4 Purification and identification of human Leukemic bone marrow lectin, using Gel filtration:

Before applying the sample to a Sephadex G-150 (1.5x75cm), the column had been equilibrated with Tris- salic buffer, pH, 7.2 containing 0.02 M CaCl₂, after that the sample was transferred at the top surface of the column and then, eluted with this buffer at an elution rate of 8 ml/ Hour. Fractions of three ml volume were collected then identified by the assay method as well as the absorbance at 280nm and protein determination were carried out. The elution volume (Ve) of the lectin in each fraction was determined by the following formula:

Ve = Fraction volume (3ml) X Fraction number containing the highest level of the lectin.

123

*Chapter Five ——— Binding characteristic studies of lectin of sera and bone marrows of Leukemic patients with Glycoprotein of human red blood cells*

## 5.7.5 The assay Methods:

In order to identify the fractions which contain lectin, the% of specific binding (SB%) for each fraction was determined as follows:

1- Half ml of each fraction isolated by gel filtration was incubated with 0.5 ml of erythrocytes in a final volume of 1ml at $37C°$ for 90 minutes, then determining absorbance at 620 nm of the upper layer (free lectin) of the assay solution.

2- Parallel experiments were performed to determine the amount of non-specific binding for each fraction, as described in section (5.6.3).

## Calculation:

The same equation mentioned in experiments (5.6.2 and 5.6.3) was used to calculate the specific non-specific binding of lectin.

124

*Chapter Five —— Binding characteristic studies of lectin of sera and bone marrows of Leukemic patients with Glycoprotein of human red blood cells*

# Results and discussion

The ability of lectin to bind the carbohydrate residues of erythrocyte surface glycoconjugates was tested using lectin from different types of leukemic and Hodgkin's patients as a source of cancerous lectin. Human blood type (B) was used in all experiment of binding studies as a source of glycoconjygates to which cancerous lectin will bind[317].

The total binding of lectin to glycoconjugates was estimated according to the homagglutination assay[317][318]. Hemagglutination is a semiquantitation procedure and has been widely used as a laboratory test because of its ease and versatility. It depends on aggregating and sedimentation of the erythrocyte after reaction with the bivalent or multivalent lectin[320].

Non- specific binding was tested using neuraminidase to be incubated with the erythrocyte suspension before the assay, this enzyme is responsible for the release of terminal sialic acid residues from the erythrocyte surface glycoconjugates, and hence, the penultimate N- acetylgulactosamine will be exposed for the lectin binding. The types of sialic acid found on the erythrocytes, give a striking correlation between the ability of lectin to agglutinate cells and the presence of sialic acid residues on mammalian cell surface[321]. This active agglutination may be due to the N- acetyl group which is present on the structure of sialo glycoproteins present on outer surface of the red cells. Many factor may influence the hemagglutination assay these include lectin amount, partial change, pH ionic strength, temperature and time of incubation[320].

125

*Chapter Five ——— Binding characteristic studies of lectin of sera and bone marrows of Leukemic patients with Glycoprotein of human red blood cells*

**Optimum conditions of the binding (Hemagglutination) of lectin (sera bone marrow) to erythycyte surface glycoconjugates**

According to the results which are shown in figures (5-1, 5-2, 5-3 and 5-4), the optimum amount of lectin which had a maximum (SB%) was $50\,\mu l$ ($200\,\mu g$) and ($3000\,\mu g$), the maximum hemagglutination activity of lectin required pH 8.5, the optimum temperature and time obtained were $37\,C^\circ$, 90 minutes respectively. From figure (5-2 and 5-3) it is clear that lectin binding is dependent on pH, temperature. The observation that percent of specific lectin binding decrease with the change of pH towards acidity or high basicity suggest that abundance of $H^+$ions in the acidic medium may inhibit the binding sites on both glycoconjugate and lectin molecules, whereas $\bar{O}H$ ions may influence the binding in the same manner, and that sialic acid which involved in binding is unstable to both acid and alkaline $pH^{(322)}$. The results indicate that the binding process is pH dependent and the shift in the pH environment may effect the stability of the macromolecules involved in the binding. This effect includes the induction of protonation- deprotonation process occurring within the ionizable groups of the amino acids present in the binding groups of these macromolecules[323].

**Effect of $Ca^{+2}$ concentration on (Hemagglutination) binding activity:**

The effect of $Ca^{+2}$ ions concentration on the binding of lectin to erythrocyte surface glycocenjugates was investigated using different concentration of $Ca^{+2}$ ions. Figure (5-5) shows the effect of $Ca^{+2}$ concentration on lectin binding, the results show that the highest binding of lectin was found in the presence of 20 mM $Ca^{+2}$ ions. The results, established, that the lectin involved in this assay is a $Ca^{+2}$ dependent, and $Ca^{+2}$ ions play

126

*Chapter Five* ——— *Binding characteristic studies of lectin of sera and bone marrows of Leukemic patients with Glycoprotein of human red blood cells*

an important role in stabilization the complex formed between the lectin and the glycoprotein present on the red cell surface. Also, the stabilization is due to the conformational changes in the protein due the binding of $Ca^{+2}$ions[324]. Different $Ca^{+2}$ dependent lectins have been purified from various sources and most of these possess multimeric structures are capable of forming cross linked complexes.

## Effect of ionic strength and different salts on the binding (Hemagglutination) activity:

The ionic strength of the incubation medium is thought to have a marked influence on the lectin binding to glycoconjugates. The effect of NaCl and $MgCl_2$ as a mono and divalent salt, on the binding of lectin to erythrocyte surface glycoconjugates was investigated. The result, show that there is a significant increase in specific binding percent in different types of leukemia and Hodgkin's disease when using different $MgCl_2$ concentrations than those obtained in the presence of different NaCl concentrations. Figure (5-6) (5-7) show the effect of NaCl and $MgCl_2$ and also the highest binding of lectin was obtained in 0.2 M NaCl and 0.015 M $MgCl_2$. The results are in agreement with the results reported by Ammar[59] and the result reported by Hassanin[204], and disagree with results reported by Nassir[58], when he has found that there is no effect of such ions on the lectin binding in the presence of 15mM $Ca^{+2}$ ions in the range of concentration used.

## Effect of denaturation agents on the binding (Hemagglutination) activity:

Different denaturating agents were used to investigate their effect on lectin binding to glycoconjugates. Table (5-1) demonstrates the effect of different concentration of urea on lectin binding. Table (5-1) shows,that the binding decreased with increasing of urea concentrations, this effect can be

**127**

*Chapter Five ——— Binding characteristic studies of lectin of sera and bone marrows of Leukemic patients with Glycoprotein of human red blood cells*

attributed to urea on the hydrophobic forces between protein molecules, also the same table shows the effect of different concentration of (PEG) on the binding[325]. Increasing concentration of (PEG) may result in precipitation of protein molecules which leads to decrease the interaction between lectin and glycoconjugates, and hence a decrease in the percent of specific binding. The effect of HCl and NaOH on the reduction of (the percent of specific binding), due to great changes in pH of incubation medium.

**Inhibition studies of the binding (hemgglutination) activity:**

The inhibition percent of binding (hemagglutination) by a number of carbohydrates (sialic acid D-glucuronicoronic acid, fructose, mannose and xylose) were studied and table (5-2) revealed that sialic acid and D-glucuronic acid were the most potent inhibitors and give high inhibition percent of hemagglutination. However, the results obtained from this assay demonstrate, that, fructose, mannose and xylose have the low activities to inhibit the binding of leukemic and Hodgkin's lectin. In a similar study it was found that 4-O, N diacetyl neuraminic acid 9,0, N-diacetglneuraminic acid are the best inhibitors for a sialic acid- binding lectin from the hemolymph of Achatina fulica snail[326].

The purification of lectin was performed in the presence of $Ca^{+2}$ions, using sephadex G-150, it has estimated the elution volume (Ve) and Kav value for elution of leukemic bone marrow lectin from sephadex G-150 and found to be 50 ml and 0.36 respectively, where as this Kav of the lectin obtained from equation (3).

CHAPTER SIX  6

# Kinetics and thermodynamic studies on binding of lectin to Glycoprotein

Lectins are carbohydrate- binding proteins and glycoproteins of non-immune origin which have been isolated from a variety of animal and plants origin[327]-[329]. They agglutinate cells or other materials that display more than one saccharide of sufficient complementarily. They are found in many categories of living things[330]. However, binding of lectins at the cell surface may cause other changes in cell function than mitogenic stimulation or the release of stimulating factor[331].

In this investigation, it is attempted to explain the mechanism of binding lectin to glycoprotein to form lectin- glycoprotein complex, and then to determine the kinetics and thermodynamic parameters, and to describe the molecular basis of lectin interaction, through the effect of time course temperature and other factors.

# Materials and methods

## 6.1 Chemicals:

All chemicals and reagents mentioned in section (2-1) of chapter two were used in the experiments of this chapter.

## 6.2 Instruments:

All instruments that used in section (2-2) of chapter two were also used in the experiments of this chapter.

## 6.3 Kinetic Studies:

**6.3.1 The time- course of cancerous leukemic lectin binding to glycoprotein present on red cell surface:**

**Reagents:**

The standard erythrocyte suspension and the assay buffer (Tris- HCl pH 8.5) containing 20 mM $CaCl_2$ and 0.2M of NaCl were prepared as described in sections (5.3 and 5.6.1) of chapter five.

**Procedure:**

1- At zero time, 0.5ml of erythrocyte suspension was added to 50 μl of purified lectin (3000 μg), the final volume of the assay mixture was made, up to 1ml with Tris- HCl buffer pH 8.5. The reaction mixture was incubated at 25 C° for several time intervals (5, 15, 30, 60, 90 and 120 minutes).

2- After each time interval, the assay tubes were treated according to step mentioned in section (5.6.2).

3- Parallel experiments were carried out according to section (5.6.3) to determine the amount of non- specific binding.

4- To determine the time- course of lectin binding to glycoprotein present on red cell surface at different temperature, the above steps were performed at (5, 15, 25, 30 and $37C^{\circ}$).

## Calculations:

1- The concentration of lectin mg/ml involved in total binding to erythrocyte surface glycoconjugates, was calculated according to the following formula:

$$\text{The concentration lectin (mg/ml) involved in total binding} = \frac{A - A^{*}}{A} \times \text{The total concentration of lectin (mg/ml) used in the assay}$$

Where:

A: The absorbance of standard erythrocyte suspension at 620 nm.

$A^{*}$: The absorbance of unbound (free) erythrocyte at 620 nm. ?.

2- The concentration of lectin in mg/ml involved in non- specific binding erythrocyte surface glycoconjugates, was calculated according to the following formula:

$$\text{The concentration of lectin(mg/ml)involved in non- specific binding} = \frac{A' - A^{*}}{A'} \times \text{total lectin concentration } (\mu u) \text{ used in the assay}$$

Where:

A' :The absorbance of neuraminidase- treated erythrocytes suspension at 620 nm.

$A^{*}$: The absorbance of unbound (free) erythrocytes at 620nm.

3- $\dfrac{\text{The concentration of specifically}}{\text{bound lectin (mg/ml)}} = \dfrac{\text{Concentration of lectin (mg/ml)}}{\text{involved in total binding}} - \dfrac{\text{Concentration of lectin (mg/ml)}}{\text{involved in non specific binding}}$

The concentration of specifically bound lectin which represents the concentration (lectin-glycojugate) complex, was expressed in micro units ($\mu u$), since $1/(\mu u)$ is the concentration of lectin (mg/ml) which gives 37% specific binding after incubation for 90 minutes at $37 C^\circ$.

4- The concentrations of specifically bound lectin (lectin glucoconjugate) complex in micro units plotted against their corresponding incubation time.

## 6.3.2 Scatchard analysis:

**Reagents:**

The Tris- HCl buffer pH 8.5 and the standard erythrocyte suspension were prepared as previously described in section (5.3 and 5.6.1) of chapter five.

**Procedure:**

1- Half ml of erythrocyte suspension was added to increasing amount of lectin (0.65- 3.3 mg/ml), the final volumes were made up to 1ml with Tris- HCl buffer (pH 8.5, contain 0.2 M NaCl and 20 mM CaCl$_2$).

2- The assay tubes were incubated for 90 minutes at $37 C^\circ$, then they treated as mentioned in steps of section (5.6.2).

3- The previous steps were repeated at different temperatures (5, 15, 25, and 30 and $37 C^\circ$).

## Calculations:

1- The concentration of specifically bound lectin ($\mu u$) was calculated for each tube according to the calculations of section (6.3.1).

2- The concentration of free (unbound or un reacted) lectin was calculated by subtracting the concentration of lectin involved in total binding from the total concentration of lectin (mg/ml) used in each experiments:

| The concentration of free lectin (mg/ml) | = | Total concentration of lectin (mg/ml) | - | The concentration of lectin (mg/ml) gives total binding |
|---|---|---|---|---|

3- The concentration of lectin binding sites ($B_{max}$) and the affinity constant ($Ka$) were determined according to scatchard equation (5):

$$\frac{B}{F} = \frac{1}{K_d} \times (B_{max} - B)$$

$$Ka = \frac{1}{K_d}$$

where:

B = The concentration of specifically bound lectin.

F = The concentration of free lectin.

Ka = The affinity constant.

$B_{max}$ = The maximal binding capacity.

$K_d$ = The dissociation constant.

4- The plot of B/F values against the values of B, gives linear relationship. The total concentration of lectin sites ($B_{max}$) was calculated from the intercept on the axis, while the value of affinity constant was calculated from the slope of the straight line.

5- The Ka and ($B_{max}$) values were also determined from the Eadie- Hofstee plot of data getting from scatchard plots, Using the following equtions.

$$B = -K_d \frac{B}{F} + B_{max}$$

The values of Ka and ($B_{max}$) were calculated from the slope of the straight line and the intercept on Y-axis respectively.

### 6.3.3 Determination of Hill-coefficient (n) of lectin binding to glycoconjugates:

**Calculation:**

1-All data obtained from the experiment in section (6.3.2) were used in this plotting method.

2-The value of Hill- coefficient (n, was calculated according to Hill equation:-

$$\log\left(\frac{B}{B_{max} - B}\right) = n \log F - \log Ka.$$

3- The values of log (B/$B_{max}$-B) were plotted against the value of log F, the Hill coefficient (n) was calculated from the slope of the straight line.

## 6.4 The thermodynamic studies:

**Procedure:**

The same steps mentioned in section (6.3.1) and section (6.3.2) were performed.

**Calculations:**

1- The thermodynamic parameters of transition state were obtained from Van'tHoff plot, the values of the natural logarithm of affinity constant (Ka) obtained at different temperatures were plotted against the reciprocal values of the absolute temperature in Kelvin (1/T), according to the following equation:

$$\ln Ka = \frac{\Delta S^\circ}{R} - \frac{\Delta H^\circ}{RT}$$

where:

$\Delta H^\circ$ : The enthalpy change of the standard state.

$\Delta S^\circ$ : The entropy change of the standard state.

$R$ : The gas constant (8.31441 J.K$^{-1}$.mol$^{-1}$)

$\Delta H^\circ$ value was obtained from slope of the linear relationship of the plot. The change in Gibbs free energy of the standard state ($\Delta G^\circ$) was obtained from the following equation:

$$\Delta G^\circ = -RT \ln Ka$$

While the entropy change of the standard state $\Delta S^\circ$ was obtained from:

$$\Delta S^\circ = \frac{\Delta H^\circ - \Delta G^\circ}{T}$$

2- The thermodynamic parameters of the transition state were obtained from Arrhenius plot of $\ln k_{+1}$ values against $1/T$ values that give a linear relationship according to the following equation:

$$\ln k_{+1} = \ln A - \left(\frac{E_a}{RT}\right)$$

where:

A = Arrhenius constant.

The value of apparent energy of activation (Ea) of the binding reaction can be determined from the slope of the straight line.

The enthalpy of the transition state $\Delta H^*$ was obtained from:

$$\Delta H^* = Ea - RT$$

The free energy change of the transition state $\Delta G^*$ is calculated from the following equation:

$$\Delta G^* = -RT \ln k_{+1} + RT \ln\left(\frac{KT}{h}\right)$$

where:

K: is Boltzmann constant ($1.38 \times 10^{-23}$ J deg$^{-1}$).

h: is Plank constant ($0.662 \times 10^{-33}$ J s$^{-1}$).

The change in entropy of the transition state $\Delta S^*$ is calculated from the following formula:

$$\Delta S^* = \frac{\Delta H^* - \Delta G^*}{T}$$

# Results and discussion:

**Kinetics of lectin binding to erythrocyte surface glycoconjugats:**

The time – course of lectin binding to erythrocyte surface glycoconjugates:

Figure (6-1) shows the time course of the formation of leukemic lectin glycoconjugate complex at five different temperatures (5, 15, 25, 30 and 37 C°). The concentration (amount) of lectin glycoconjugate complex that formed after time (t) was calculated from the following equation:

| The concentration of (lectin- glycoconjugate) complex formed after time(t) (mg/ml) | = | The concentration of lectin involved in total binding (mg/ml) | - | The concentration of lectin involved in non- specific binding (mg/ml) |
|---|---|---|---|---|

- The concentration of (Lectin- glycoconjugate) complex formed after time (t) was expressed in ($\mu u$) since 1 ($\mu u$) is the concentration of lectin mg/ml which gives 37% specific binding at 37 C° and 90 minutes.

The results of time- course pattern at different temperature indicate that the lectin (obtained from sera) binding to erythrocyte surface glycoconjugates is a temperature and time dependent process, and the maximum binding can be obtained at 37 C° after incubation for 90 minutes.

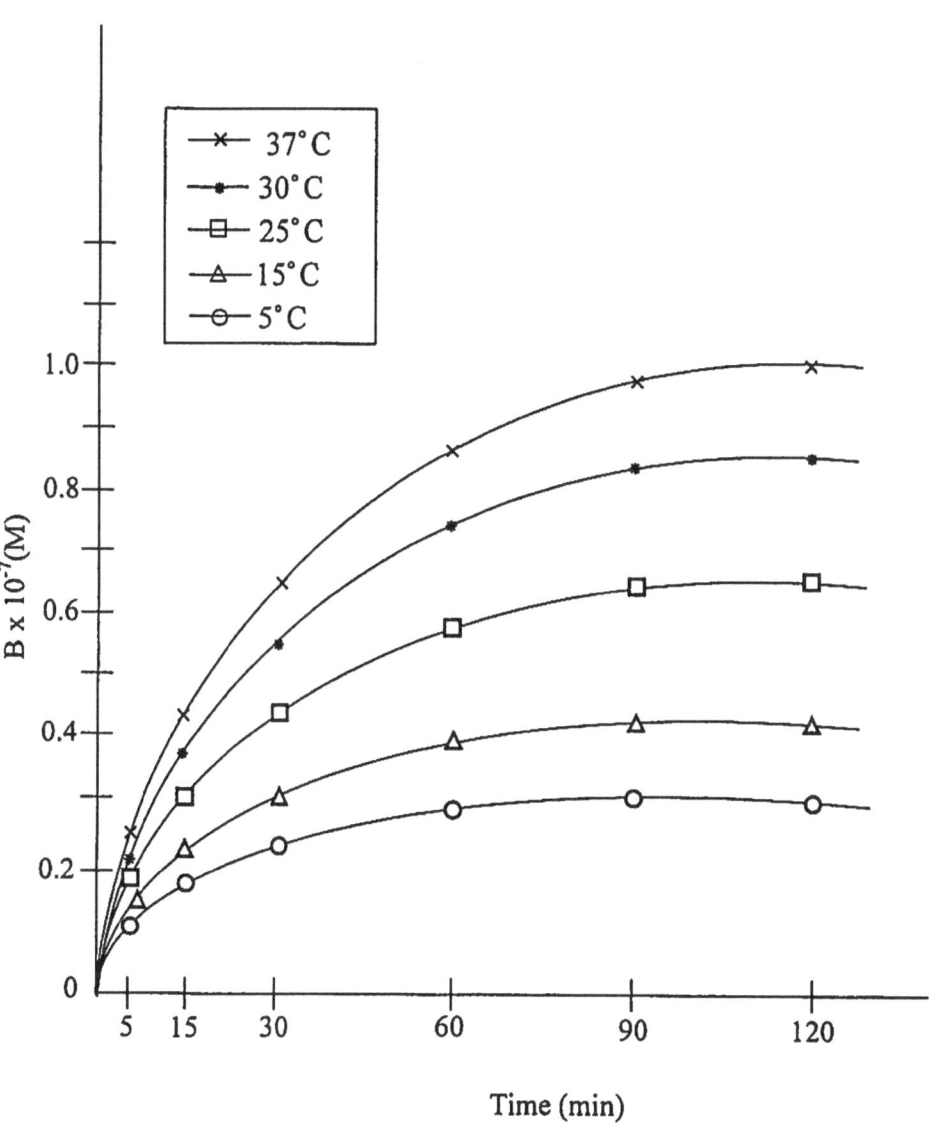

**Figure (6-1)**

Time-course of lectin binding to erythrocyte surface to

glycoconjugates at different fife temperatures.

All details are explained in the text.

**Determination of kinetic parameters of lectin- glycoconjugate complex formation:**

The time- course of lectin binding to erythrocyte surface glycoconjugates was performed to describe the kinetic parameters of binding. The simplest proposed model representing the binding of lectin to glycoconjugates could be expressed by the following equation:

$$\text{Lectin} + G \underset{K_{-1}}{\overset{K_{+1}}{\rightleftharpoons}} \text{Lectin} - G \qquad \text{....(1)}$$

where:

G: glycoconjugate

$K_{+1}$: is the rate association of lectin with glycoconjugate.

$K_{-1}$: is the rate dissociation of formed under the same conditions.

At equilibrium:

$$Ka = \frac{(\text{Lectin} - G)}{(\text{Lectin})(G)} \qquad \text{......(2)}$$

$$K_d = \frac{(\text{Lectin})(G)}{(\text{Lectin} - G)} \qquad \text{.....(3)}$$

Thus:

$$Ka = \frac{1}{K_d} = \frac{K_{+1}}{K_{-1}} \qquad \text{.....(4)}$$

where:

$Ka$ = is the equilibrium constant of the association (affinity constant).

$K_d$ = is the equilibrium constant of the dissociation of lectin-G complex.

Table (6-1): shows the values of Ka, Kd and maximal binding capacity of lectin ($B_{max}$), which was calculated from Scathard and Eadie- Hofstee plot (Figures 6-2 and 6-3) respectively at different temperatures (5, 15, 25, 30 and $37 C°$).

The $K_d$ values were calculated employing equation (4).

It is clear from table (6-1), that the affinity constant (Ka) is temperature dependent parameter (Ka increase from $5 \times 10^6$ $U^{-1}$ at $5 C°$ to $9.6 \times 10^6 U^{-1}$ at $37 C°$). This change indicate that the lectin exhibits the maximum affinity to bind erythrocyte surface glycoconjugates at $37 C°$ after incubation 90 minutes, and the lowest $K_d$ value of lectin – glycoconjugate complex occurs at $37 C°$ after incubation for 90 minutes.

**Table (6-1) The kinetic parameters of binding to erythrocyte surface glycoconjugate. All details are explained in the text**

| $T C°$ | $Ka \times 10^6 U^{-1}$ | $K_d \times 10^{-7}$ (U) | $B_{max} \times 10^{-7}$ (U) |
|---|---|---|---|
| 5 | 5 | 2 | 7 |
| 15 | 6.1 | 1.6 | 7.7 |
| 25 | 7.5 | 1.33 | 9.3 |
| 30 | 8.37 | 1.19 | 10.2 |
| 37 | 9.64 | 1.03 | 11.4 |

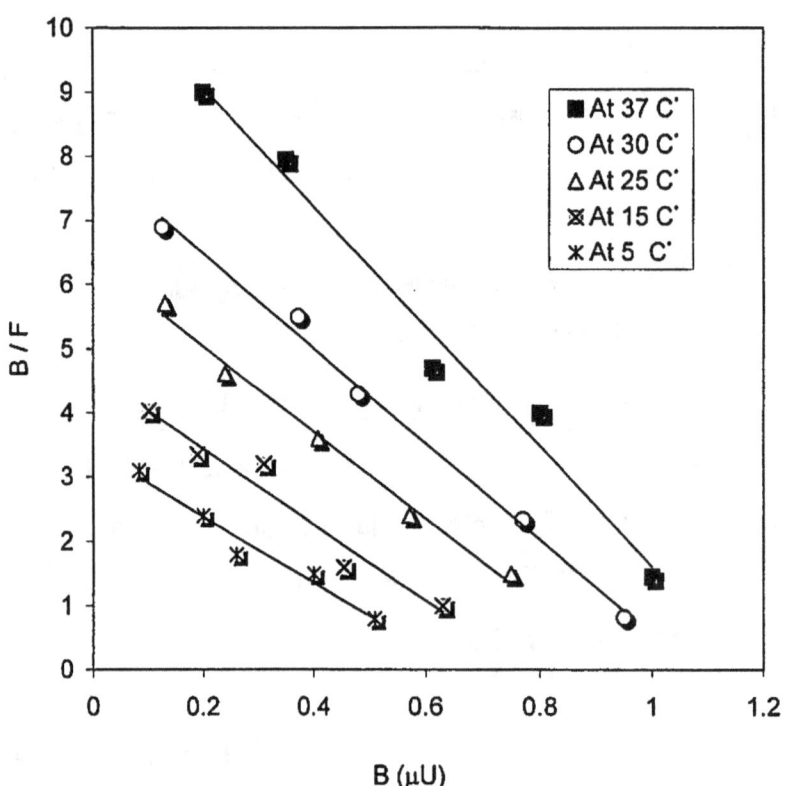

Figure (6-2)
Scatchard plot for the binding of leukemic lectin to
erthrocyte surface glycoconjugates.
All other details are explained in text.

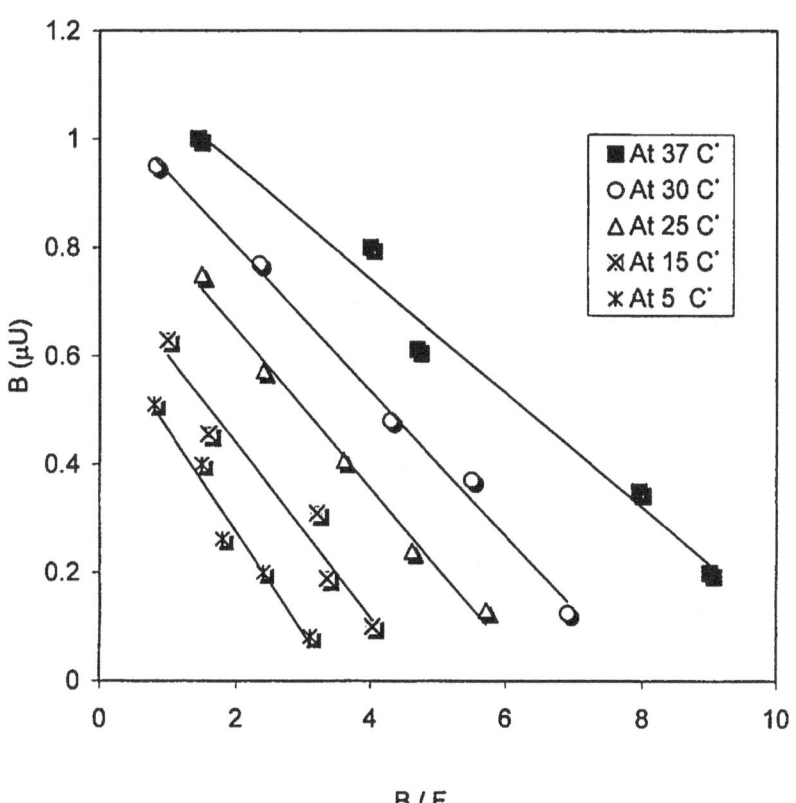

Figure (6-3)
Eadie-Hofstee plot for the binding of leukemic lectin to
erthrocyte surface glycoconjugates.
All other details are explained in text.

However, the time- course data obtained from figure (6-1), could be used to determine the reaction order of lectin binding to erythrocyte surface glycoconjugates using the following equation in

$$\ln[\text{lectin} - G]_e \left[ \frac{(\text{lectin})_t - (\text{lectin} - G)_t (\text{lectin} - (G)_e /(G)_t}{(\text{lectin})_t [(\text{lectin} - G)_e - (\text{lectin} - G)_t]} \right] =$$

$$K_{+1} t \left[ \frac{(\text{lectin})_t (G)t}{(\text{lectin} - G)_e} - (\text{lectin} - G)_e \right] \qquad \qquad ....(5)$$

where:

$K_{+1}$: is the kinetic association constant in $U^{-1} min^{-1}$.

$(\text{lectin})_t$: is the concentration of lectin at time (t).

$(G)_t$ : is the concentration of glycoconjugate at time (t)

$(\text{lectin-G})_t$ : is the concentration of lectin- glycoconjugate complex at time (t).

$(\text{lectin-G})_e$ : is the concentration of lectin- glycoconjugate complex at equilibrium.

Equation (5) represents the second order kinetics. But in our work in which the percent of specific binding was, in some cases, small and most of the lectin remains free and only small fraction binds even at equilibrium, ie, $(\text{lectin})_t \gg (\text{lectin-G})_e$ thus,

$$(\text{Lectin})_t \gg \frac{(\text{Lectin} - G)_t (\text{lectin} - G)_e}{(G)_t} \quad \text{and} \quad \frac{(\text{lectin})_t (G)_t}{(\text{lectin} - G)_e} \gg (\text{lectin} - G)_e$$

so that the following equation could be used in order to fit the pseudo first order kinetics:

$$\ln \frac{(\text{Lectin}-G)_e}{(\text{lectin}-G)_e - (\text{Lectin}-G)_t} = K_{+1}t \frac{(\text{lectin})_t(G)_t}{(\text{lectin}-G)_e} \qquad \ldots(6)$$

On the other hand, Figure (6-4) shows the plot of $\ln \dfrac{(\text{Lectin}-G)_e}{(\text{lectin}-G)_e - (\text{Lectin}-G)_t}$ against time (t) gives a straight line with slop equal to the observed value of first order constant $(K_{obs})$ in $\text{min}^{-1}$. The rate constant $(K_{+1})$ was calculated at five different temperatures by using the following equation:

$$K_{obs} = K_{+1} \frac{(\text{lectin})_t(G)t}{(\text{lectin}-G)_e}$$

$$\therefore K_{obs} = K_{+1}(\text{lectin})_t \qquad \ldots(7)$$

The values of $K_1$ at five different temperatures were calculated by using equation (4). Also the half life time of association (t 1/2)$_{ass}$, which represents the time needed for the formation of half amount of the complex at equilibrium, was determined from the concentration of the complex at equilibrium and the time- course curve. While the half life time of dissociation (t 1/2)$_{diss}$, was calculated from the following relation :

$$(t_{1/2})_{diss} = \ln \frac{2}{k_{-1}} = \frac{0.693}{k_{-1}} \qquad \ldots(8)$$

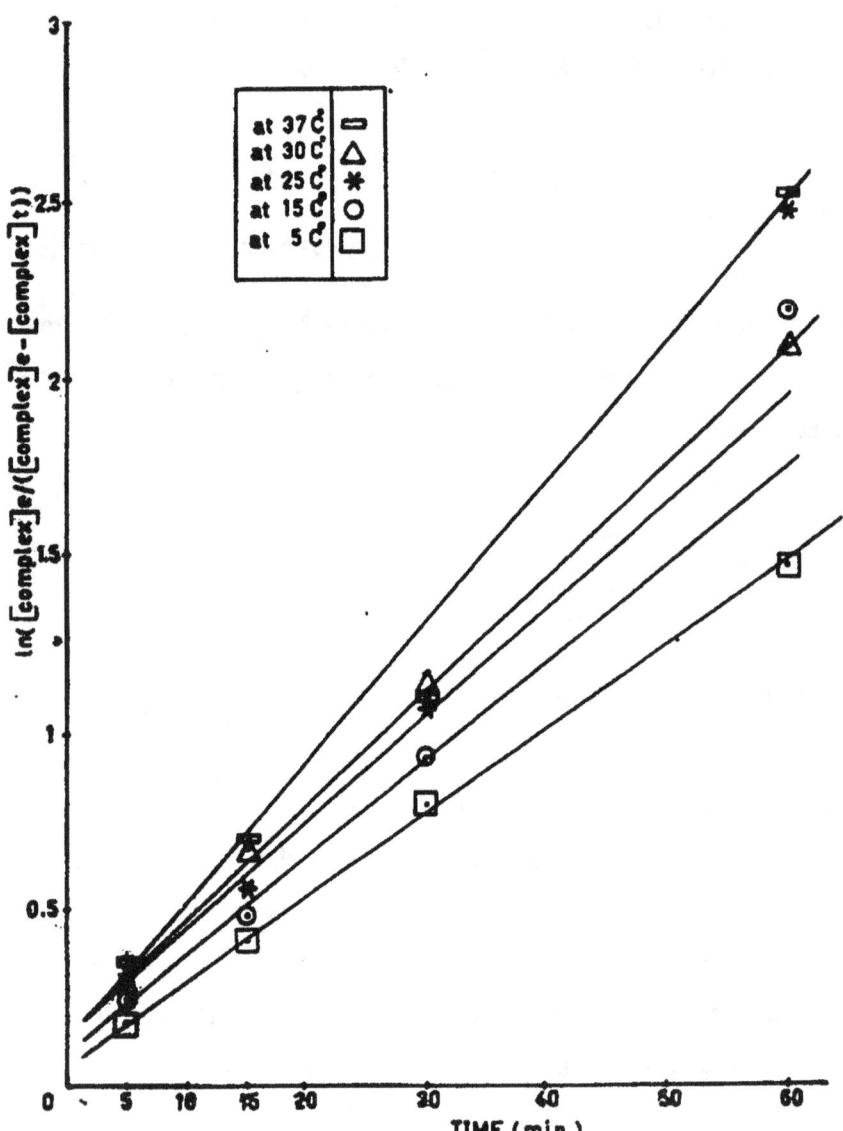

**Figure (6-4)**
Kinetics of lectin binding to erythrocyte surface to
glycoconjugate at different five temperatures.
All details are explained in the text.

The values of $k_{obs}$, $k_{+1}$, $k_{-1}$, $(t_{\frac{1}{2}})$ ass, and $(t_{\frac{1}{2}})$ diss at five different temperatures are summarized in table (6-2). The results revealed that the association rate constant $(k_{+1})$ at $37\,C^\circ$ is the highest one among other values at $5\,C^\circ$, 15, 25, and $30\,C^\circ$. Also the result show that the dissociation rate constant $(k_{-1})$ glycoconjugate complex depends on temperature.

**Table (8-2): The effect of temperature on the kinetic parameters of lectin binding to erythrocyte surface glycoconjugates. All details are described in the text**

| Temperature $C^\circ$ | $k_{obs}$ $(min^{-1})$ | $k_{+1} \times 10^3$ $(Umin^{-1})$ | $k_{-1} \times 10^{-3}$ $(min^{-1})$ | $t_{1/2ass}$ $(min)$ | $t_{\frac{1}{2}diss}$ $(min)$ |
|---|---|---|---|---|---|
| 5 | 0.025 | 12.1 | 2.42 | 28 | 286 |
| 15 | 0.035 | 16.9 | 2.77 | 22 | 250 |
| 25 | 0.040 | 19.4 | 2.58 | 18 | 268 |
| 30 | 0.042 | 20.4 | 2.4 | 15 | 288 |
| 37 | 0.0453 | 22.0 | 2.28 | 14 | 303 |

**Scatchard Analysis:**

Figure (6-2) shows Scatchard plot of lectin binding to erythrocyte surface to glycoconjugate at different five temperature (5, 15, 25,30 and $37\,C^\circ$) after incubation time 90 minutes. This figure could be used to determine kinetic parameter of lectin binding, such as, the equilibrium constant of dissociation of complex $(K_d)$ and total concentration of lectin binding sites $(B_{max})$ of human leukemia's lectin by using the following equation (5):

The values of the parameters at these different temperatures are summarized in table (6-1). The result, shows that the concentration of lectin binding sites ($B_{max}$) is temperature dependent explained according to the number of molecules possessing the activation energy for interaction, increase with increase temperature. On the other hand, the affinity constant (Ka) is also depended on temperature, this indicates that the reaction is slightly endothermic and explained by the fact that affinities of endothermic reactions enhanced by increasing temperatures. However, the values of $B_{max}$ and $K_d$ for human leukemia's lectin at different temperatures obtained from Scatchard analysis were similar to those obtained from the Eadie- Hoststee plot Figure (6-3)[332].

**Determination of Hill-coefficient (n) of lectin binding to glyconconjugates:**

Figure (6-5) represent Hill plot of lectin binding to erythrocyte surface glycoconjugate in five different temperatures (5, 15, 25, 30 and 37 C°), the value of Hill-coefficient (n) equals the slop of resulting straight line. The values were obtained at 5, 15, 25 30 and 37 C° and they were (1.4, 1.3, 1.035, 1.07, 0.99) respectively. The cooperatively of the lectin binding sites could be estimated through the determination of Hill-coefficient (n). The results obtained in this work indicates that the cooperatively of lectin binding sites was low affected by temperature.

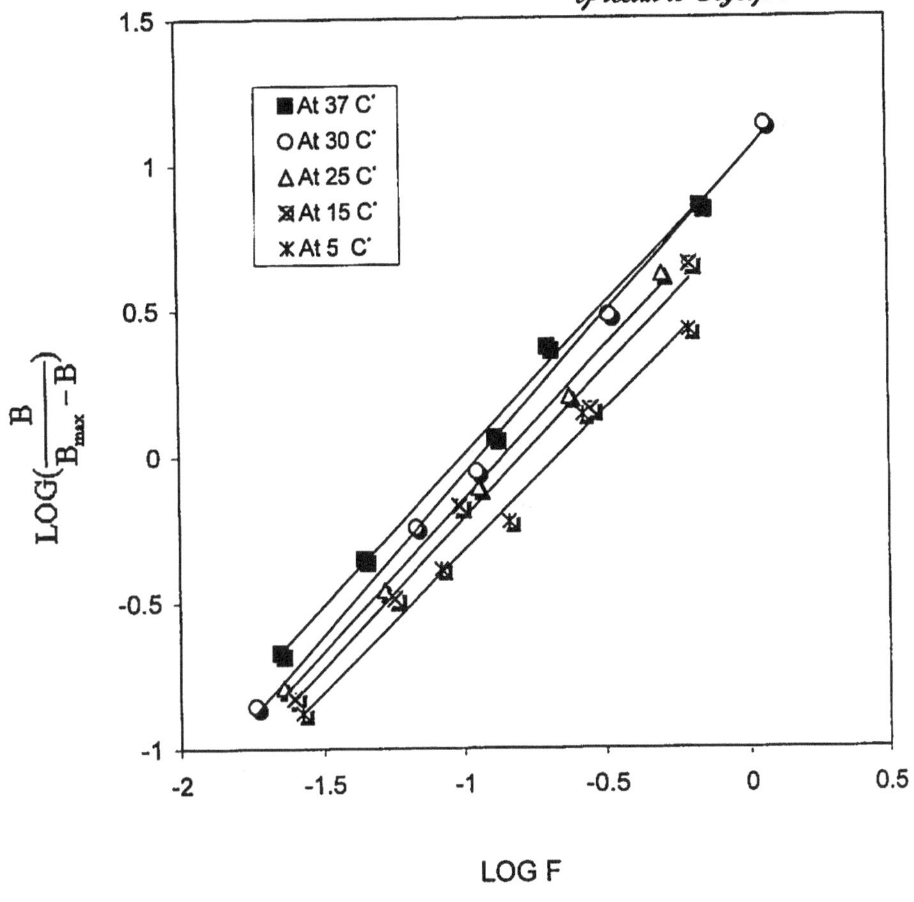

Figure (6.5)
Determination of Hill-coefficient (n) of lectin binding to
erthocyte surfuce glycoconjugates at four temperatures,
A- 5 C˙, B- 15 C˙, C- 25 C˙, D- 30 C˙ and E- 37 C˙
All other details are explained in text.

**The thermodynamic of the lectin binding to erythrocyte surface glycoconjugates:**

**Thermodynamic parameters of standard state:**

The dependence of the equilibrium binding constant (affinity constant) for the binding of lectin to erythrocyte surface glycoconjugates on temperature can be observed from van't Hoff plot figure (6-6).

The result, obtained from van't Hoff plot revealed that $\Delta H°$ in general had a positive value of 14.72 KJ/mol., and that the reaction was nearly endothermic. The small positive value of $\Delta H°$ may indicate a fovorable interaction between the lectin and glycoconjugate sub groups. These include the non-covalent interaction which are fundamentally electrostatic in nature such as charge- charge, charge- dipole, dipole- dipole, charge- include dipole, dipole- include dipole, and hydrogen bond. The sum of these types of interaction can yield some stabilization to the folded structure of the complex.. Table (6-3) shows the values of $\Delta G°$ of five temperatures (5, 15, 25, 30, and 37 C°). The results revealed that the $\Delta G°$ values increases with decreasing temperature, since it, value was- 41.44 KJ/mol at 5 C°, -40.15 KJ/mol at 15 C°, -39-21 KJ/mol at 25 C°, -37.40 KJ/mol at 30 C° and -35.69 KJ/mol at 37 C°, it can be said that the lectin binding erythrocyte surface glycoconjugates needs higher energy at low temperatures, also the negative values of $\Delta G°$ indicates the stability of lectin- glycoconjugate complex, subsequently the high affinity of the reactant. The high negative values of $\Delta G°$ indicates that the binding of lectin to glycoconjugates is spontaneous reaction. In addition, these values are controlled by high positive $\Delta S°$ values, table (6-3). The result, showsthat the values of $\Delta S°$ decrease with increasing temperature to the more stable and more arranged status of lectin glycoconjugate complex at $37\,C^{\circ}$, high values of positive $\Delta S^{\circ}$ suggest that the binding spontaneity was entropically driven. Entropy was the driven force for the occurrence of the binding, this indicates that the hydrophobic interactions played an important role in stabilizing the complex.

Table (6-3): Thermodynamic parameters at standard state of lectin binding to erythrocytes glycoconjugates.

"All details are explained in the text"

| Temperature $C^{\circ}$ | $\Delta H^{\circ}$ (KJ/mol) | $\Delta G^{\circ}$ (KJ/mol) | $\Delta S^{\circ}$ (J/mol.K) |
|---|---|---|---|
| 5 | 14.727 | - 35.64 | 181.1 |
| 15 | 14.727 | - 37.40 | 181.0 |
| 25 | 14.727 | - 39.21 | 181.0 |
| 30 | 14.727 | - 40.15 | 181.1 |
| 37 | 14.727 | - 41.44 | 181.1 |

**Thermodynamic parameters of transition state:**

According to the transition state theory the interaction of two proteins leads to the formation of an activated complex (transition state), consequently, the interaction of lectin with erythrocyte surface glycoconjugates can be represented as follows:

Lectin+ glycoconjugate → [lectin-glycoconjugate] → lectin- glycoconjugate
                              an activated complex        (final product)
                              (transtion state)

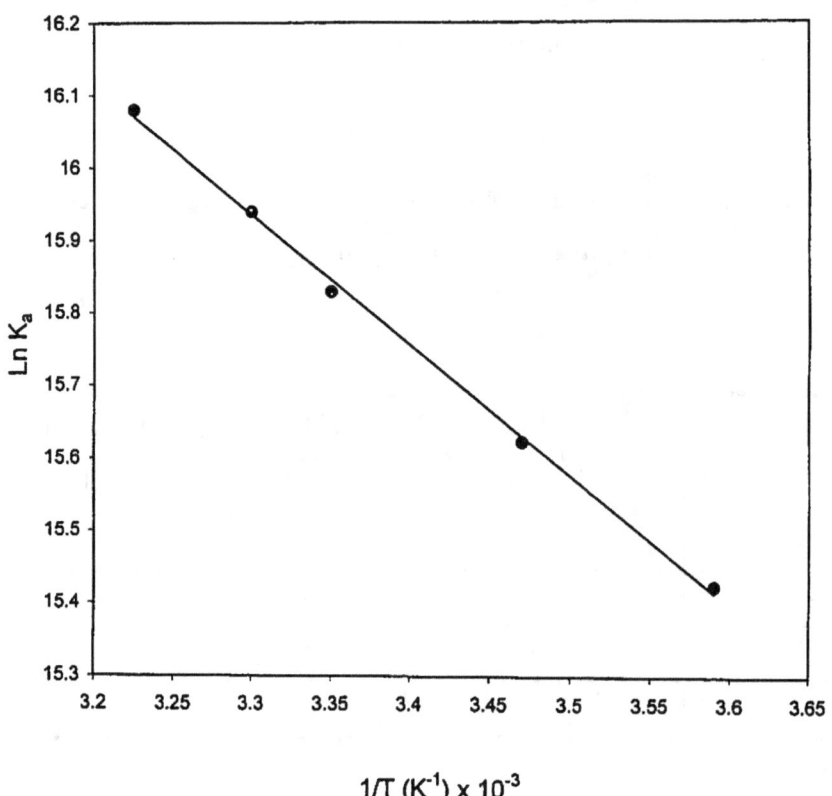

Figure (6-6)
Van't Hoff plot for the binding of leukemic lectin to
erthrocyte surface glycoconjugates.
All other details are explained in text.

On application of Arrhenius equation to the kinetic data, we could determine the transition state thermodynamic parameters ($\Delta H^*, \Delta G^*$ and $\Delta S^*$) at five different temperatures (5, 15, 25, 30 and 37 C°).

Figure (6-7) shows Arrhenius plot of $\ln k_{+1}$ against $1/T$ values. The slope of the straight line represents the activation energy (Ea).

The values of the thermodynamic parameters of the transition state (Ea, $\Delta H^*, \Delta G^*$ and $\Delta S^*$) are summarized in table (6-4). The high value of activation energy (8.5 KJ/mol) represent the required energy to overcome the energy barrier of the transtion state for the formation of lectin- glycoconjugate complex.

However, the value of activation energy is accordance with the positive values of $\Delta G^*$, indicates that the formation of an activated complex non-spontaneous process.

The results, in table (6-4) show the values of $\Delta H^*$ at five different temperatures (5, 15, 25, 30 and 37 C°), and revealed that the $\Delta H^*$ values decreased with increasing temperature.

The slight changes in the values of $\Delta H^*$ at different temperatures could be attributed to the dependence on $\Delta H^*$ an activation energy Ea, through the equation:

$$\Delta H^* = Ea - RT$$

Since the numerical value of RT is too small in comparison with the value of activation energy for the binding lectin to glycoconjugates.

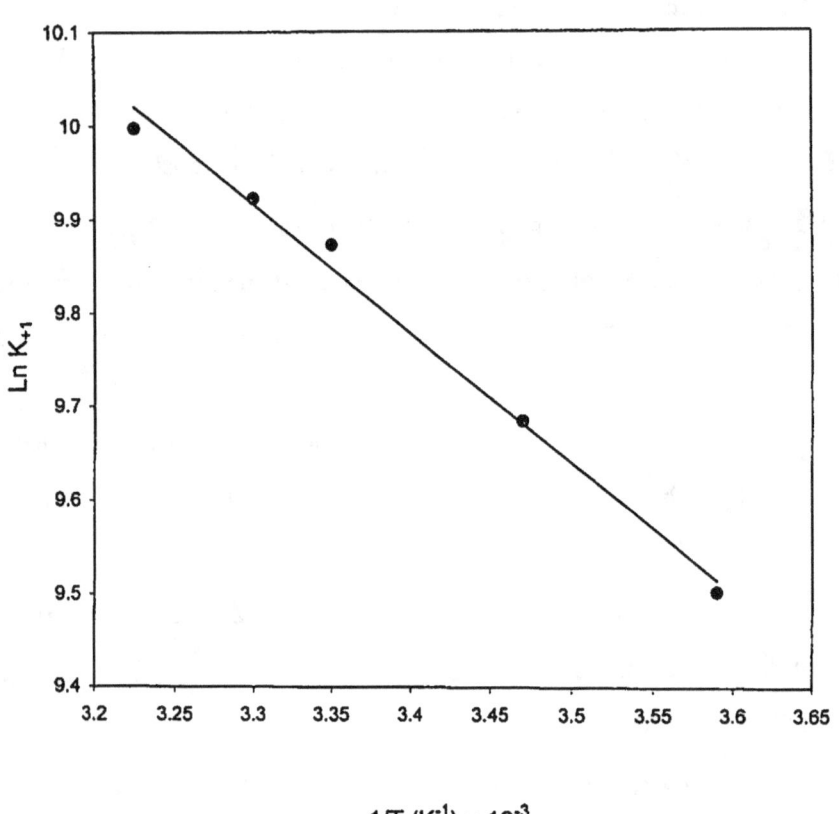

Figure (6-7)
Arrhenius plot for the binding of leukemic lectin to
erthrocyte surface glycoconjugates.
All other details are explained in text.

**Table (6-4): Thermodynamic parameters at transition state of lectin binding to erythrocyte surface glycoconjugates.**

**"All details are explained in the text"**

| T C° | Ea (KJ/mol) | $\Delta H^6$ (KJ/mol) | $\Delta G^6$ (KJ/mol) | $\Delta S^°$ (J/mol.K) |
|------|------------|------------|------------|------------|
| 5 | 8.5 | 6.188 | 47.40 | - 148.24 |
| 15 | 8.5 | 6.10 | 48.38 | - 146.8 |
| 25 | 8.5 | 6.021 | 49.7 | - 146.57 |
| 30 | 8.5 | 5.98 | 50.41 | - 146.66 |
| 37 | 8.5 | 5.922 | 51.38 | - 146.63 |

Table (6-4) shows the increasing of $\Delta S^*$ values with the elevation of temperature $\Delta S^*$ values were- 148.24 J/mol. K at 5C°, -146.8 J/mol.K at 15C°, - 146.57 J/mol.K at 25C°, -146.66 J/mole.K at 30C° and -146.83 J/mol.K at 37C°.

The high negative values of $\Delta S^*$ indicate that the activated complex (lectin-G) involved in the binding process had a more arranged structure than the starting reactants (lectin and glycoconjugate). Finally, it could be conclude that the values of the thermodynamic parameters obtained from the study of lectin binding to erythrocyte surface glycoconjugates, give a distinct idea about the nature of force that regulate the formation of the complex.

In order to compare the values of transition state with those of standard state, it is suggested to have the thermodynamic model to describe the formation of the complex.

This model is illustrated in fig (6-8). The thermodynamic model proposes that the formation of lectin- glycoconjugate complex undergoes three thermodynamic states. The thermodynamic state A, represent the initial energy level of lectin and glycoconjugate. The thermodynamic state B, represents the association of the two species to form the activated complex (lectin- glycoconjugate). The thermodynamic state, C, represents the complete binding of lectin with glycoconjugate and formation of the complex.

The model involves two steps, the reaction at step 1 is associated with positive $\Delta G^*$ value, this indicates that the binding of lectin to glycoconjugates in this step requires external energy. Also in step 1, the lectin binding shows negative value for entropy change ($\Delta S^*$), this negatively indicates the alteration in the structure of lectin- glycoconjugate transition complex to arranged one. At step 2, the contribution of the activated complex in more interactions, giving the fully interacting complex (lectin-G).

The formation of protein- ligand complex is proposed to occur in two steps, the first is, the stabilization of the complex by hydrophobic interaction and the second is the stabilization by short range interactions, such as electrostatic interactions, protonation, hydrogen bonding and van der waals interaction [333].

Hydrophobic interactions contribute to the complex stability via high positive entropy ($\Delta S^* > 0$), whereas, the electrostatic interactions, and Van der waals inter actions contribute to the stability of the lectin- glycoconjugate complex via negative entropy change ($\Delta S^* < 0$) [333][334].

The thermodynamic data from the present study indicate that the binding of lectin to erythrocyte surface glycoconjugates is entropy driven and is in agreement with concept that hydrophobic interactions play an important role in such reactions.

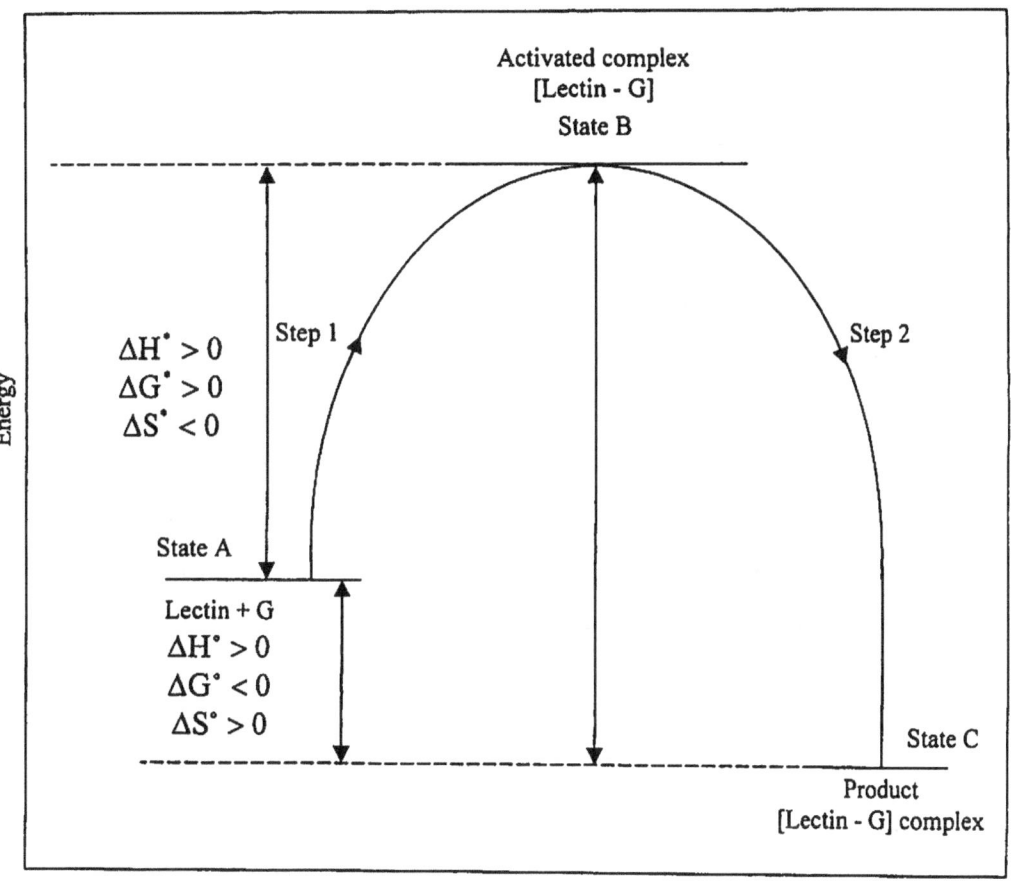

**Figure (6-8)**
General energy diagram and thermodynamic model applied to the complex formation between lectin and erythrocyte surface lycoconjugates

# General discussion

The result in chapter two revealed that the means of TP concentration show a slight change in leukemia and Hodgkin's patients. The result revealed also an increase in TSA and LASA levels in all patients leukemia, Hodgkin's and control in comparison to those of normal individuals.

There was a highly significant difference between the TSA and LASA compared to normal and control. The specificity and sensitivity of TSA and LASA were examined. The TSA/TP ratio showed a significant difference between leukemia, Hodgkin's to normal subject.

The TSA, LASA and TSA/TP ratio were elevated in serum of patients of leukemia, which had no response to chemotherapy and before death. These results showed that TSA, LASA and TSA/TP ratio could be useful as bio chemical marker for diagnosis, monitoring and staging of disease.

A significant elevation were also observed in the levels of mucoid protein in serum of all types of leukemia and Hodgkin's disease, compared to normal and control patients. Also the level of protein- bound hexoses in sera of patients of leukemia and Hodgkin's were significantly increased compared to those of normal and control. From these results, it is possible to conclude that sero mucoid protein and protein- bound hexose were useful clinical marker for differential, diagnosis and monitoring of leukemia and Hodgkin's disease.

Lactate dehydrogenase levels were elevated significantly in all groups of leukemia and Hodgkin's patients compared to normal. It is observed that LDH activity is found to be in-patients, which had no response during the course of chemotherapy. On the other hand the values of SOD activity decreased in leukemia and Hodgkin's patients sera compared to that normal and control individuals.

The values of serum SOD activity obtained from this assay were correlated with LASA concentrations in sera of leukemia, Hodgkin's and control patients.

The results disclosed that there is a significant negative correlation between the two parameters.

The determination of trace elements and electrolytes shows that there is elevation in Cu level Cu/Zn ration in sera of leukemia and Hodgkin's patients. On the other hand there is decreased of Zn levels and the levels of Ca and Mg remain in normal level, in sera of leukemia and Hodgkin's patients.

The results also revealed that the highest specific binding of lectin from sera and bone marrow of leukemia and Hodgkin's disease was 38% at pH 8.5, for 90 minutes, at $37 C^\circ$, in the presence of 20 mM of $Ca^{+2}$ ions. The binding seemed to be effected by ionic strength. The lectin was purified from human leukemic bone marrow by using of gel filtration method.

The kinetic parameter of lectin binding such as, the equilibrium constant of dissociation of complex ($K_d$) and total concentration of lectin binding sites ($B_{max}$) of human leukemia lectin. The results show that the concentration of binding sites ($B_{max}$) and the affinity constant (Ka) is temperature dependent, this indicate that the reaction is slightly endothermic. The cooperatively of the lectin binding sites could be estimated through the determination of Hill- coefficient (n). The results indicate that the cooperatively of lectin binding sites was low effected by temperature and the result from van't Haff plot revealed that the $\Delta H^\circ$ in general had positive value and the reaction were nearly endothermic. The value of $\Delta G^\circ$ of five temperatures (5, 15, 25, 30 and $37 C^\circ$) increase with decreasing temperature, this means that the lectin binding erythrocyte surface glycoconjugates need higher energy and the negative values of $\Delta G^{\circ}$ indicate that the binding of lectin to glyconjugates is spontaneous reaction. In addition these values are controlled by high positive $\Delta S^{\circ}$ values, and the $\Delta S^{\circ}$ decrease with increasing of temperature to more stable status of lectin glycoconjugate complex at 37 C°.

From Arrhenius equation the thermodynamic parameters for the transition state determined at five different temperatures ($\Delta H^{*}$, $\Delta G^{*}$ and $\Delta S^{*}$) the high activation energy represent, and positive values of $\Delta G^{*}$ indicate that the activated complex non- spontaneous process. $\Delta H^{*}$ decrease with increasing temperature and $\Delta S^{*}$ elevation with increasing temperature.

The thermodynamic data from the present study indicate that the binding of lectin to erythrocyte surface glycoconjugates is entropy driven.

**Conclusions**

1- Total sialic acid (TSA) and lipid associated sialic acid (LASA) measurements appear to have high sensitivity for wide range of tumors such as Leukemia and Hodgkin's and other cancers.

2- The level of protein-bound hexose and seromucoid was increased in sera of Leukemic and Hodgkin's patients and other cancers, but it may not specific enough to be a good diagnostic test for these cancers, but they could be useful with other diagnostic tests for diagnosis and monitoring Leukemic and Hodgkin's patients.

3- The superoxid dismutase activity and SOD/LASA ratio appear to be putative markers of Leukemia and Hodgkin's disease with potential use fulness.

4- The results indicate that lectin may recognize the sialyl residue of the glycoprotein and the binding is due to sugar specific inter action.

5- The kinetic study of lectin binding to glycoconjugates revealed that the binding reaction is a time and temperature dependent process.

## The future works

1- Measurement of TSA, LASA and LASA/TP ratio in sera of patients/pre and post treatment to monitor its activity and stage disease.

2- Estimation the levels of sialic acid in the CSF of patients with leukemia and Hodgkin's disease.

3- Combination of TSA with other markers to improve the specificity of TSA as a tool for the diagnosis of human leukemia and other cancer.

4- Purification of lectin from sera and bone marrow of patients with leukemia and Hodgkin's disease and do spectra scopy studies (IR, NMR) to them.

5- Estimation the SOD activity in CSF of leukemia and Hodgkin's disease and correlate the activity with other parameter (marker) such as LDH, sialyl transferase.

1- Edwards, C.R.W. and Bouchier, I.A.D. (1995) Davidson's principles and practice of MEdicine. 17$^{th}$ Ed chapter 13, pp 803-815.

2- Firkin, F. and Chesterman, C. (1989) Clinical Haematology in medical practice. 5th Ed. Black well. Scientific publication. Chapter 10. p.236-276.

3- Beck, W.S. (1991) Hematology 5th Ed. lecturer, 22 p. 399-417.

4- Borson, R. MD and Loeb, V.Jr. (1994) "Adult leukemias, Vol.44, No.6, November/ December (p.386-349).

5- Haff brand, A. V and Lewis, S. M (1999) Postgraduate Haematolgy 4th Ed. chap. 18. P. 373-40.

6- Pierce, J. H. and Eva, A. (1986). clin Haemat, 15: 573.

7- Kurzock. R. and Gutteman, J. U. N. (1988), Eng. J Med. 319: 990-998.

8- West brook CA. et al. (1992) Blood, 80: 2983-2990.

9- Eanelios, G.P. et al. (1983) Hematol. 20, 1.

10- Perdersen- Bjergaard, J. et al. (1982) Lancet. 11. 83.

11- Jacobs, A. (1989) Br. J. Haematol. 72: 119-121.

12- Hayes. R.B. et al. (1997) J. Nath. Cancer inst. 89, 1065-1071.

13- Weiss, R.A. and Marshall, C.A. (1984) Oncogenes Lancet, ii, 1138.

14- Evan, G.I. and Lennox, E. S. (1985). Brit. Med. Bull. 41, 59.

15- Green, P.L and Chen, I. SY. (1990) FASEB. 4: 162-175.

16- Huang, M.Li, C.Y. et al. (1984), Blood. b4, 42.

17- Bennett, J.M. and Catovsky, D. D. (1976) Br. J. Hemat: 33, 451.

18- Bennett, J.M. and Catovsky, D. D. (1982) Brit. J. Hemat: 51, 189.

19- Bennett. J.M. et al (1988). Ann. Int. MEd. 103,460.

20- Hoelzer, D. (1994) Semin Hematol; 31: 1-15.

21- Quesada, JR. et al. (1984), N. Eng. J. Med. 310: 15-18.

*References* ————————————————————————

22-  Santos, G.W. (1989) Blood; 74: 901-908.

23-  Santos, G.W (1990) Cancer; 65 suppl; 786-791.

24-  Linker, CA. (1992), Cum opinion oncol. 4: 53-65.

25-  Estey. E. and Thall, P. (1994). J. Clin oncol, 12; 671-678.

26-  Burrentt, A.K. and Eden, O. B. (1997). Lancet; 849: 270-708.

27-  Burrett, A.K. and Sterens, R. (1988). Lancet; 351: 700-708.

28-  Sell, S. (1991); Human. patho; 21: 1003-1019.

29-  Burtis. C. A. and Ash wood, E. R. (1994), TIETZ. Textbook of clinical chemistry secth adition chap 5. p 897.

30-  Vaitukaitis J.I. (1979). Ahor mone secretEd of or mangreason. 301-326.

31-  Chu. T.M. (1992). The Humana press; pp.99-115.

32-  Sell, S: Cancer markers In: comprehensive (1991), Textbook of oncology. 2nd Ed. Vol. J.A.R. Moossq, et al. pp.225-238.

33-  Kiang, D.T. Greenberg, L.J. (1990); Cancer. 65: 193-99.

34-  De Bie, S.H., Ferreria, J.C. (1992); J. cancer Res. clin. oncol., 118: 1-15.

35-  Burtis. C. A. and Ash wood, E. R. (1994) TIETZ. Textbook of clinical chemistry 2Ed chapter 4.

36-  News Landes clinical applications of tumor markers. (1987); Med lab Sci. 44: 361-370.

37-  Chu. Tn. (1987) curr Top path; 77: 19-45.

38-  Torosian, MH. (1988) Surgy Gynecol obes. 166: 567-579.

39-  Patel, P.S. and Baxi, B. R. (1991) Int. J. Biol. marker; 6(3): 177-182.

40-  Patel, P.S. and Baxi, B. R. (1997) Indian. J. Bichem, Biophys. 34: 226-233.

41-  Ganz, P. A. (1987) Clin. oncol. 5: 472-9.

42-  Pui, C. H. (1988) Blood; 66: 778-782.

*References* —

43- Nathanson, L. and Fishman, W. H. (1971) Cancer (phila). 27: 1388-97.

44- Nouwen EJ cancer Res. (1985); 45: 892-902.

45- Van Balgooy, JNA. (1979) Comp biochem physiol; 62B: 263-268.

46- Oberley, L.W. (1979) cancer Res. 39: 1141-9.

47- Yoshimitsu, K. and Usui, T. (1984) Acta paediatr scand. 73: 42-46.

48- Borrello, S. et al. (1992) FEBS Lett; 310:249-254.

49- Johnson, PWM. (1993) Br J. Cancer; 67: 760-66.

50- Toshihiko Myrayama. et al. (1997) Int J. Cancer; 70: 575-81.

51- Patel, P. S. et al., (1995) Neoplasma; 42(5): 271-76.

52- Oztokatli, A. et al. (1992) Int. Urol. Nephrol; 24: 125-129.

53- Yue, K. et al. (1995) Acta Acad Sci; 17(2): 128-31.

54- Patel, P.S. et al. (1993) Gynecol oncol; 50(3): 294-301.

55- Gatchev. O. and Rastam, L. (1993) Br. J. Cancer; 68: 425-27.

56- Vadralova, E. et al. And Borovarsky, J. (1994) Cancer Lett; 78: 171-172.

57- Riley, M. and Taut, C. (1990) Clin chem; 36: 161-163.

58- Nassir, M.F. phd thesis. (1995) Submitted to the College of Science, Baghdad University.

59- Ammar, K.G., phd thesis. (1999) Submitted to the College of Science, Baghdad University.

60- Dwived, C.M. and Hardy, R. E. (1990)Experientia; 46: 91-94.

61- Seider, A. et al. (1992) Pad. Undpadol; 27:43.

62- Patel PS. Adhvaryu SG, Balader, DB. (1988) Tumori; 74: 639-644.

63- OkennEdy, R. and Berns, G. (1991) Cancer lett; 58: 91-100.

64- Schamberger, R.J.J. (1984) Clin. Biochem; 22: 647-51.

65- Voigtmann, R., Pokorny, (1989) J. Cancer; 64: 2279-83.

66- Patel, P.S. et al. (1994) Anti- Cancer. Res; 14: 747-57.

67- Silver, H,K.B. (1979) Cancer Res; 39: 503-642.

*References* ─────────────────────────────

68- Silver, HKB. (1981) Surg Gynecol. obes; 153: 209-13.

69- Berra B. Rapeli, (1986) Int J. Biol markers; 1: 39-46.

70- Ozhiganov, EL. et al. (1987) Lab Delo; 3: 204-7.

71- Tautu, C. et al. (1988) J. Nath can Insl; 80: 1333-37.

72- Li, R. X, Ladisch. (1991) Biochem Biophys Acta; 1046: 57-65.

73- Polivkova, J. et al. (1992) Neoplasma; 34(4): 233-6.

74- Potoukalian. J. et al. (1993) Int J. Cancer; 53: 948-51.

75- Wong, Y.F. et al. (1993) Med Sci. Res; 21: 397-8.

76- Dunzendor fer, U. et al. (1981) Urol; 194.

77- Dnistrian, A.M. et al. (1982) Cancer; 50: 1815.

78- Murayama, J. et al. (1997) Anti cancer Res; 17(4A): 2545-48.

79- Tomas ZewskA, R. et al. (1997) ActaopaEd Jap; 39:448-450.

80- Romppannen, J. et al. (1998) Anti, cancer. Res; 18(4B): 2793-2797.

81- Trabelsi, N. et al. (1997) Environ- Health. Respect; 5:1153-8.

82- Burger, AM. et al. (1997) Clin. Cancer. Res; 3(3): 455-63.

83- Sinha, D. and Mandal, C. (1999) Leukemia; 13(1): 119-25.

84- Sinha, D. and Bhattacharya, D. K. (1999) Leu. Res; 23(5): 433-9.

85- Nigam VN, Cantero. Adv. (1973) Cancer. Res; 17: 1-80.

86- Ajit Varki and Sandra Diaz, (1983) J. Biol., Chem; 258: 12465-12471.

87- Alido, L. et al. (1995) Clinica Acta. 243: 165-179.

88- Boucheir, I.A.D., and Clamp. J.R (1971). Clinichim. Acta; 35: 219-224.

89- Sheshadri Warayana. (1994) Ann. Clin. Lab. Sci; 24(4): 376-384.

90- Schauer, R. (1982) Adv. Carbohydr. Chem. Bio chem; 40: 131-239.

91- Schauer, R. (1991) Glyco biology. 1: 449-452.

92- Varki, A. (1992) Glycobiology. 2: 25-40.

93- Edelman, G. M. (1983) Science. 219: 450-451.

94- Miguel, A. Ferrero. et al. (1996) Biochemj. 317: 157-165.

95- Miburn, J. (1981) J. Biol. chem. l: 264: 487.

*References* _____

96- Hakomari, S. (1974) Adv. cancer. Res. 18: 265-315.

97- Lipton, A. Harvey, HA. (1979), Cancer. 43: 1766-71.

98- Bergelson. L.D. et al. (1982) Eur. J. Biochem. 128:467-474.

99- Rosherg, SA. Einstein, AB. (1972) J. Cell Biol 53: 466.

100- Kamerling, J.P. Schauer, AR. (1982) Biochem. Biophy. Acta. 714: 351-355.

101- Paulson, J.C. (1985). in the receptors (conn, P.M., Ed.) Vol.11, pp.131-219 Academic press, orland, F1.

102- Suzki, Y. et al. (1985) J. Biol chem. 266: 1362-65.

103- Marsh., M. (1989) Adv. Virus. Res. 36: 107-115.

104- Hoekstra, D. and Kok. JW. (1989) Bio Sci. Rep. 9:273-305.

105- Schauer. R. Trends. (1985) Biochem. Sci. 10: 357-60.

106- Freagkiadakis, GA. Stratakis, EK. (1997) Comp. Biochemphysiol. 117B: 545.

107- Shi, WX. et al. (1996) J.Biol. chem. 271: 31526.

108- Schauer, R. (1983) Biochem Soc Trans. 11: 270.

109- Varki, A. J. et al. (1991) cell. 65:65-74.

110- Varki, A. J. (1980) Exp Med. 152-532-544.

111- Maria, C. P.L. et al. (1995) Biochem Biophy. Acta. 236: 323-330.

112- Crook. M. (1993) Clin Biochem. 26:31-38.

113- Ravindaranath, M.H. Paulson, J.C. (1988) J Biolchem. 263:2079-2086.

114- Sen G, Chowdhury, M. Mandal, C. (1994) Mol cell Biochem. 136: 65-70.

115- Mandal, C. Sinha, D. (1997) Ind J. Biochem Biophys. 34: 82-86.

116- Hakomori, S.F. (1985) Cancer Res. 45: 2405-14.

117- Warren, L. Fuhrer. J.P. (1972) Proc. Natl. Acadsci.

118- Svenner bolm. L. (1957) Biochem Biophy Acta. 24: 604-11.

119- Warren, L. (1959) J. Biol chem. 234: 1971-75.

*References*

120- Shukla, AK. and Schauer, R. Hoppe. Seyler's Z. (1981) Physiol. Chem. 362: 236-237.

121- Shamberger, R.J. (1968) Anti cancer Res. 6: 717-20.

122- Hara, S. Takemori, Y. et al. (1989) Anal. Biochem. 164: 138-145.

123- Bossmann, H. Hall, T. Proc. (1974) Natl Acad, Sci. USA. 71:1833.

124- Xing. RD. Chen, R. M. (1994) Int J. Biol. Markers. 9(4): 239-42.

125- Dnistrian, AM. Schwartz, MK. (1983) Ann- clin. Lab. Sci. 13: 137-42.

126- Schutter, EMJ. et al. (1992) Tumor Biol. 13: 121-132.

127- Svennerblom, L. and FreEdman, RA. (1980) Bio Chem. Biophy Acta. 617: 79-109.

128- Prinstrian, A.M. Schwartz, M.K. (1981) Clin chem. 27: 1737.

129- Stringou, E. et al. (1992) Anti cancer Res. 12: 251-255.

130- Patel. P.S. Baxi, B. R. (1989) Neoplasma. 36(1): 53-59.

131- Pigman, W. (1977). The glycoconjugates 1 stEd Vol.1, Academic press, pp.181-188.

132- Martin, D.W.J. (1985) Harper's review of bio chemistry.

133- Filipe. M.I. and Fengar, C. (1979) Histo chem . J. 11, 277-287.

134- Glick, M.C. and Flowers, H. (1978). The glycoconjugates 1 stEd. Vol.2, Academic press, pp.337-384.

135- Thomas. M. Devlin (1986) Text Book of Biochemistry with clinical correlation 2nd P (102-103).

136- Holden, K.G. et al. (1987) "Respiratory tract" in the glycoconjugates" Ed. Horwitz M. I. and Pigman W. 1stEd. Vol.1, Academic press, pp.215-237.

137- Irwin, J. Goldstein. et al. (1980) Nature, 28 (5): 66.

138- Dipti. G. and FrEd. C; Bewer, (1994). J. Bio chemistry. 5526-5530.

139- Nathan Sharon , et al. (1972) Science September; 177: 4053: 949-959.

140- Lis, H. and Sharon., N. (1986). Ann Rev. Biochem. 55:35-67.

*References* ——————————————————————

141- Yamamato, K., et al. (1981) Biochem. J. 195: 701-713.

142- Cummings, R.D. and Kornfeld. S. (1982) J. Biol. chem. 257:11235-
      11240.

143- Tollefesen, S. E. and Kronfeld, S. (1983) J. Biol. chem. 258:5165-5171.

144- Peters, B.P. et al. (1979) Biochemi, J. 18: 5505-5511.

145- Roche, A.C. et al. (1975) FEBS Lett. 57:245-249.

146- Mohan, et al., (1982) Biochem. J. 203: 253-261.

147- Babal, P. (1994) Biochem. J. 299(2): 341.

148- Kawagishi, H. (1944) FEBS Lett 340:56.

149- Cohen, E. (1984) "Recognition Proteins, Recepters and probes:
      invertebrates progress in clinical and biological research, Vol.
      157. New York; Lis, pp.207.

150- Gilbsa- Garber, N. et al. (1985) FEBS Lett. 81: 267-270.

151- Vasta, G.R., Cheng. et al. (1984) J. Cell Immunol. 88: 475-488.

152- Goldstein, I. and Hayes, C. (1978) Adv. carbohydrate. chem. Biochem.
      35, 127.

153- Miller, J. et al. (1975) Proc. Natl Acad. Sci. USA 72. 4236.

154- Beyer., E. et al. (1980) J. Biol. chem. 255, 2536.

155- Ceri, H. et al. (1981) J. Biol. chem. 256: 390-94.

156- Kornfield, R. and Korn field, S. (1970) J. Biol. chem. 245, 2536.

157- Barondes, S. Annu- Rev. (1981) Bio chem. 50: 207.

158- Cheresh, D.A. et al. (1984) J. Biol chem. 259: 7453-7459.

159- Basu, S. Sarkar, M.Mandal. C. (1986) Mol cell Biochem. 71: 149-157.

160- Crocker, P.R. Klems. et al., (1991) EMBO. 10: 1661-69.

161- Fukuda, M. et al., (1985) J. Biol chem. 260: 12957-12967.

162- Fukuda, M. et al. (1986) J. Biol chem. 261: 12796-12806.

163- Smith BA, Ware BR. (1978) J. Immunol. 120: 921-925.

164- Boldt, DH. et al. (1975) J. Immunol. 115: 1532-36.

*References* ————————————————————————

165- Scott, R.E. and Rosenthal, A.S. (1977) J. Immunol. 9: 37-47.

166- Mridula, C. et al. (1985) Bio chem. Biophy. Res. commun. 136: 1301.

167- Cohsl. Lincolnst. et al. (1986) Cancer Invest. 4: 305-327.

168- Plucinsky. M.C. et al. (1986) cancer. 56: 2680-2685.

169- Khanderia, U. et al. (1983) J. surg. oncol. 23(3): 163-6.

170- Patel, P.S. et al. (1990) Anti cancer Res. 10:1071-1074.

171- Hakomori, S.I. (1981) Ann. Rev. Biochem. 50:733-736.

172- Bossman, G.J. et al. (1982) Biochem. Biophys. Acta. 22:693(2): 444-450.

173- O-KennEdy, R. et al. (1982) Eur. J. Cancer. clin. oncol. 18(5): 437-444.

174- Hakomori, S.I. et al. (1984) Ann. Rev. Immunol. 38: 289.

175- Silver, H. Murray. R. (1983) Int. J. cancer. 31: 39.

176- Raynes. J.G. Biom, Ed. (1983) pharmocother. 37(3): 136-8.

177- Erbil, K. et al. (1985) Cancer. 55: 404.

178- Horgan, I. (1982) clin. chem. Acta. 118: 327.

179- Gail, H.M. et al. (1986) JNCI. 76: 805.

180- Kim, Y. and I saacs. (1975) Cancer Res. 53: 2097.

181- Alhadeff, J.A. (1984) Crit. Rev. oncol. Hematol. 9(1): 37-107.

182- Michael, A.B. et al. (1985) Blood. 66(5): 1068-1071.

183- Cohen, A.M., et al. (1989) Eur. J. Haematol. 42: 200-203.

184- Fabio. Pio. Nadia. M. et al. (1989) Int. J. Cancer . 44: 434-439.

185- Rothenberg, RE. et al. (1994) Breast. Dis. 7,3: 197-202.

186- Vegh Z.S. et al. (1991) Clin. Chem. Acta. 203: 259-268.

187- Lopez, J.J.B. and Serna, A.V. (1995) Int. J. Biol. Marker. 10(3): 174-179.

188- Lowry, O.H., Rosebroug, N.J. et al. (1951) J. Biol. chem. 93: 265.

189- Katapodis, N. et al. (1982) Cancer. Res. 42: 5270-75.

*References*

190- Wilknson, L. Systat. (1990). The system for statistic Evanston, II: systatinc.

191- Hoff brand, A.V. and pettit, J.E. (1985) Essiental Hematology . 2Ed. Black well.

192- Warren, L. Buck CA. et al. (1978) Biochem. Bio phy. Acta. 516: 97-127.

193- Patel, PS. Baxi, B. r (1990) Indian journal pathol. Microbiol. 33: 124-128.

194- Baxi, Br. Patel, P.S. et al. (1991) Cancer. 67: 135-140.

195- Sherblow, A.P. et al. (1980) J. Biolchem. 225: 783.

196- Yogess warnac. (1983) Adv. cancer. Res. 38: 289-295.

197- Yogess Warnac and salk, P.L. (1981) Science. 212: 1514-16.

198- Petru, E. Sevin, Bu. et al. (1990) Gynecol. oncol. 38: 181-6.

199- Adamo, V. Altavilla, G. et al. (1985) Miner VamEd. 76: 2067.

200- Schulzec. Arch. (1990) Geschwvl. Sto forsch. 66: 19.

201- Simms, Z. et al. (1979) Gim. Pol. 50:957.

202- Dimisttian, Am. et al. (1982) Cancer. 50:1815-1819.

203- Black, PH. N. (1983) Engl. J. MEd. 303: 1415-1419.

204- Hassanin, K, B. MSC thesis. (1999) Submitted to college of Science Mustansria University.

205- Reinlgein, D.S. et al. (1992) An plastsurg. 28: 55-59.

206- Mannello, F. et al. (1993) Breast cancer. Res. Treat. 24: 167-170.

207- Haq. M. Haqs. et al. (1993) Ann. clin. Biochem. 30:283-6.

208- Towmbis. et al. (1992) Anti. Cancer. Res. 46(2): 157-162.

209- Turner. C.A. et al. (1985) J. Clin. Pathol. 38: 588-592.

210- Sakai. T. et al. (1990) clin chem. 36:474-476.

211- Adelbert, S.B. et al. (1974) Cancer. Res. 34:538-542.

212- Walker. C. and Gray. (1983) Cancer. 52:150-154.

*References* ─────────────────────────────────

213- Stringou, E. Chom dor SK. et al. (1992) Anti cancer. Res. 12: 251-6.

214- Harshman, S. et al. (1974) Cancer. 34:291.

215- Weimer, and Mashin. (1965) J. clin chem principle and techniques.

216- Bradley, W.P. et al. (1977) Cancer. 40: 2264-2272.

217- Bhuvarah amurthy, V. et al. (1992) Biochem. Int. 28:105.

218- Yamamooto, K. et al. (1984) Eur. J. Biochem. 143:133.

219- Blomer. S. and Davidson , E. (1981) Biochem. 20:1047.

220- Yaskhika, T. watar u. I. et al. (1988) Am.J. Nephrol. 2:21.

221- Leninger, A. N. D. & Cox, M. (1993) Principle of bio chemistry 2nd
        Ed., pp.416-437.

222- Kaplan, A. (1989) Clinical chemistry, 2nd Ed., pp.784-93.

223- Burtis. C, A. and Ashwood, E. R. (1994) TIETZ Text book of clinical
        chemistry 6th Ed. chap 18. pp. 812-818.

224- Schneider, RJ. et al. (1980) Cancer. 46:139.

225- Ridgway, D. et al. (1981) J. PEdiatr. 99:611.

226- Endrizzi, I. et al. (1982) Eur. J. cancer, clin-oncol. 18:945.

227- Hagberg, H. (1983) scand. J. Haematol. 13:19.

228- Koziner, B. et al. (1984) Cancer. 53: 3692.

229- Flanagen, N.G. et al. (1989) Clin-Lab. Haematol. 11(1): 17-20.

230- Bergier, I. et al. (1990) Pol. Aroh. MEd. Wew N. 83(4-6): 161-165.

231- Drey fuss, A.I. et al. (1992) Cancer. 15:70(10): 2499.

232- Lanter, A.L. (1975) Clinical Biochemistry. 6th Ed. chapter.17.pp.557.

233- Fasolo, G. et al. (1989) Int.J. Biol. marker. July; 4(3): 142-8.

234- Fanin, R. et al. (1989) Haematologice. 74(2): 161-5.

235- Pandit, M.K. et al. (1990) Indian. J. Pathol. Microbiol. 33(1): 41-47.

236- Korin thenberg, R. et al. (1990) Acta pacdiatr- scand. 79(3): 335-42.

237- Hanada. S. et al. (1997) Cancer. chemother- pharmacol. 40:547-50.

238- Brambilla, P.G. et al. (1997) Minerva. Med. Jul-Aug; 88(7-8): 311-316.

*References*

239- Shimazaki, C. et al. (1997) Intr- J. Hematol. Jul; 66(1): 111-115.

240- Herndon, J. E. et al. (1998) Chest. Mar. 113(2): 723-731.

241- Gacouin, A. et al. (1998) Intensive. care. Med. Mar. 24(3): 265-7.

242- Samha, H. et al. (1998) Leukemia. Aug: 12(8):1281-2.

243- Suzuki, M. et al. (1998) Leuk. Lymphoma. Feb; 28:583-90.

244- Ferrara, F. et al. (1998) Leuk. Lymphoma. Apr; (29): 375-82.

245- Oda, N. Nakai, A. et al. (1998) Eur. J. Endocrinol. 139(3): 323-9.

246- Milone, J. et al. (1998) Bome- marrow. Transplantation. Nov; 22(10): 1019-1021.

247- Baker. K.S. Gordon, B.G. et al. (1999) J. clin. oncol. Mar; 17(3): 825-31.

248- Bindi, M. et al. (1999) Recenti. Prog. Med. Apr; 90(4): 213-5.

249- Dumontet, C. et al. (1999) Leukemia. 13(5): 811-7.

250- Karan, P. et al. (1999) Cas. Lek. Cesk. Jan.; 18(2): 40-6.

251- Howell, B., M. S. & Schaffer, R. (1979) Clin. chem. 24: 828-830.

252- Freer, D. Statland, B. et al. (1979) Clin. chem. 25: 565-569.

253- Wacker, W. D. & Vallec, B. N. (1956) J. Med. 255: 449-456.

254- Mista, I. (1972) J. Biol. chem. 247: 6960-6962.

255- Wever, R. (1973) Biochem. Biophys. Acta. 302: 475-8.

256- FEder, M. et al. (1975) J. Lab. Clin. Med. Feb.; 337-41.

257- Suresh, A. et al. (1994) Cancer. Gene. Ther. Jun; (2): 85-90.

258- Maneva, A. et al. (1995) Eur. J. cancer. per vertion. 4: 429-435.

259- Kinight, JA. (1995) Ann. clin. Lab. Sci. 25:111-121.

260- Jenner, P. (1994) Lancet. 344: 796-798.

261- Andersen, H.R. et al. (1997) clin chem. 43(4): 562-68.

262- Fridorich, I. (1972) Acc chem. Res. 5: 321-326.

263- Wever, R. et al. (1973) Biochem, Biophys. Acta. 302: 475-478.

264- Mc cord, J. M. and Fridovich, I. (1969) J. Biol. chem. 244: 6049.

_References_ ———————————————————————————

265- Batalic, R. Klein, B. et al. (1989) Clin. Exp. Heamatol. 7: 319-328.

266- Siguresson. et al. (1992) N. eng. J. Med. 326:363-67.

267- Borrello, S. et al. (1992) Neoplasma. 39(4): 233-236.

268- Van Balgooy, J. MA. and Robert, E. (1979) Comp biochembiophys. Acta. 62(B): 263-268.

269- Gorman, A. et al. (1997) FEBS Lett. Mar 3; 404(1): 27-33.

270- Kobayashi. D. et al. (1997) Blood. Apr (1); 89(7): 2472-9.

271- Kurokawa, M. et al. (1997) Anti. cancer. Res. Jul-Aug; 17(4A): 2545-8.

272- Kumerova. AO. et al. (1997) Eksp. Klin. Farmakol. 60(3): 48-50.

273- Gross, A. et al. (1998) Free. radical. Res. Feb; 28(2): 179-91.

274- Watanabe, N. (1998) Leuk. Lymphoma. Aug; 30(5-6): 477-82.

275- Ma. Y. Cao. L. (1998) Free. radic. Biol. Med. Sep; 25(4-5): 568-75.

276- Giri. DK. et al. (1998) J. Immunol. Nov1; 161(9): 4834-41.

277- Yabuki. M. et al. (1999) Free. radic. Biol. Med. Feb; 26(3-4): 325-32.

278- Kumerova, A. et al. (1999) Mater. Med. Pol. Jan- Jun; 30(1-2): 12-5.

279- Hildeman, DA. et al. (1999) Immunity. Jun; 25(3): 735-744.

280- Lui. GY. et al. (1999) Mol. Carcinog. Jul; 10(6): 196-206.

281- Burtis, C.A. & Ashwood, E.R. (1994) TIETZ. Text book of clinical chemistry 2th. chapter. 28. pp.(1317-1353).

282- Frieden, E. (1984) Biochemistry of essential ultratraceelements. New York, Plenum press. Chapter 28, pp.1-15.

283- Mertz, W. (1981) Science. 1 213-1332.

284- Sunderman, F.W. (1973) Hum pathol. 4: 549.

285- Sunderman, F.W. (1975) Ann. Clin. Lab. Sci. 5: 421.

286- Burtis, C.A. Ashwood, E.R. (1994) TIETZ Text book of clinical chemistry 2th. Chapter (29), pp.(1354-1374).

287- Winterboum, C.C. Hawkins, R.E. (1975) Lab. Clin. Med. 85(2): 337-341.

*References*

288- Passey. R. B. (1973) post grad Med. 53:173.

289- Sarvanan, C.S. et al. (1977) Indian. J. cancer. 14:38.

290- Roa, Y.N. et al. (1978) Indian. J. cancer. 5:39.

291- Bhatnagar, A. et al. (1983) Indian. J. Med. Res. 78: 127.

292- Al- Mudaffar, S. A. and Rassam, M.B. (1979) Biochem Appl. Biol. 15(3): 238-244.

293- Pui, C.H. Dodge, F.K. et al. (1985) Blood. 66: 778-782.

294- Heinoren, P.K. et al. (1987) Tumori. 73: 301-302.

295- Flanagan, N.G. et al. (1989) Clin Lab. Haemat. 11:17-26.

296- Knee, J.K. Mitidieri, E. (1991) Cancer Lett. 57: 199-202.

297- Bolzan, A.D. Bianchi, M.S. (1993) cancer. Res. 6: 142-6.

298- Fernandez. Pol., J. A. et al. (1987) Cancer. Res. 42:604-617.

299- Larry, W. oberley and carry, R.B. (1979) Cancer. Res. 39: 1141-1149.

300- Galeotti, T. Mosotti. et al. (1991) Xenobiotica. 21(8): 1041-51.

301- Oberley, L. & Oberley, T.D. (1988) Mol. cell. Biochem. 84: 147-153.

302- Wong, Y.F. Wong, W.S.H. et al. (1993) Med, Sci. Res. 21:397-398.

303- Al-Samaraee, E.H. (1997). Thesis. Submitted University of Baghdad. Science collage.

304- Abdullah, B. Dashti, Hyat. L. et al. J.N: (1989) metabolism of minerals and race element in human disease. Abdullah M. Dashti, H. Sarker, B. et al. Editors. Smith. Gordon and Company limited.

305- Abdullah, B. Dashti, H. and Mathur, M. (1985) In Element in health and disease. said, H. Editor. karachi, Hamdard. pp.197.

306- Toms G.C. and western. A. (1971) Acandemic press, New York. Chapter 10, pp.367-463.

307- Watkins, W.M. and morgan. J. W. T. (1951) Nature. 57: 359.

308- Well, P. N. (1964) Cancer. Res. 20: 462.

309- Goldstein, I.J. Hollerman. C.E. (1965) Bio chemistry. 4: 876.

*References* ————————————————————————

310- Bishayee, S. and Dori, D. (1980) Biochem. Biophys, Acta. 623: 89.

311- Bohiool, B.B., Schmidt, E. L. (1975) J. Bacteriol. 125: 1188.

312- Barger, M. and Goldberg, A. (1976) Proc. Nat Acod Sci, 125: 1188.

313- Nowak, T. and Barondes, S. (1975) Bio chemica, e+. Bio, phys. Acta. 393: 15.

314- Miller, et al. (1987) Methods in Enzymology. Vol. 138: pp. 527-530.

315- Liener, I. (1955) Arch. Biochem. Biophys. 54: 223.

316- F,S. Parker. (1971) Applications of INFRA RED spectroscopy, in biochemistry, biology and medicine. Plenum press. New York. Chapter (18) pp. 440.

317- Lis, H. and Sharron, N. (1977) Methods in Enzymology. Vol. 28: pp.360.

318- Kaplan, A., Li, S. and Kehoe, (1977) Biochemistry. 16: 42-97.

319- Scopes, R. (1982) Protein purification principles and practice, springer verlag. pp. 162, New York , Heidelber Berlin.

320- Kaplan, A. (1998) Clinical chemistry, theory, Analysis and corelation, 2Ed. pp. 180.

321- Shulka, A.K., and Schauer, R. Hoppe. (1982) Seyler's and physiol. chem. 363: 255-262.

322- Gott Shalk, A. (1960) The chemistry and biolog of sialic acids and relation substances, Cambridge University press, London.

323- Baktiear, M. Ph. D. Thesis, (1992), Submitted College of Science, Baghdad University.

324- Wild, J., Robinson, D. (1983) Biochem. J. 21: 167.

325- Dipti, G., FrEd, C. and Brewer, (1994) J. Biochemistry. 33: 5526-30.

326- Chitra, M. and Sujata, B. (1987) Biochem. Biophys. Res. Common. 148(2): 795-801.

327- Lis, H, and sharon, N. (1973) Annual Review of Biochemistry. 42: 541.

*References* ————————————————————————

328- Noriko, T. et al. (1981) J. Bio. chem. 256: 5345.

329- Vdo, S. Hans, P. et al. (1989) Histo chemical, J. 21: 44.

330- Goldstein, I.J. et al. (1980) Nature. 285: 66.

331- Scatchard, G. Ann. N.Y. (1949) Acad Sci. 51: 660.

332- Emil, L. In (1985) principles of biochemistry, seven Ed. pp.289, 107, 624.

333- Blumenthar, D.K and Stull, J. T. (1982) Biochemistry. 21: 2386.

334- Laport, D.C. et al. (1980) Biochemistry. 19: 3814.

*References* _____

328- Noriko, T. et al. (1981) J. Bio. chem. 256: 5345.

329- Vdo, S. Hans, P. et al. (1989) Histo chemical, J. 21: 44.

330- Goldstein, I.J. et al. (1980) Nature. 285: 66.

331- Scatchard, G. Ann. N.Y. (1949) Acad Sci. 51: 660.

332- Emil, L. In (1985) principles of biochemistry, seven Ed. pp.289, 107, 624.

333- Blumenthar, D.K and Stull, J. T. (1982) Biochemistry. 21: 2386.

334- Laport, D.C. et al. (1980) Biochemistry. 19: 3814.

www.ingramcontent.com/pod-product-compliance
Lightning Source LLC
Chambersburg PA
CBHW080807180526
45168CB00006B/2355